Practice Multiple-choice Papers
suitable for:

Key Skills Level 2 Application of Number
and
Level 2 Adult Numeracy

Elizabeth Jones

Published by
Lexden Publishing Ltd
www.lexden-publishing.co.uk

Acknowledgement:

Thanks are due to Roslyn Whitley Willis for her expert advice and guidance regarding the Key Skills standards and for her constant support and encouragement during the development and completion of this resource book.

First Published in 2008 by Lexden Publishing Ltd.

ISBN: 978-1-904995-51-7

Lexden Publishing Ltd

Email: info@lexden-publishing.co.uk
www.lexden-publishing.co.uk

Printed by Lightning Source

Contents

Introduction

This book provides a resource that can be used by both tutors who deliver, and students who are following Key Skills Level 2 Application of Number and Level 2 Adult Numeracy.

The difference between a Key Skills qualification and an Adult Numeracy qualification

A Key Skills qualification is achieved by candidates who pass the externally-set and examined End Assessment paper and submit a successful Portfolio of Evidence.

An Adult Numeracy qualification is awarded to candidates who pass the End Assessment test, referred to as the National Adult Numeracy test.

The End Assessment question paper for both qualifications is the same.

What does an End Assessment for Key Skills and Adult Numeracy examine?

The external assessment of both courses is in the form of a test containing 40 multiple-choice questions. These papers aim to assess the candidates' ability to read and understand numbers, graphs, charts and diagrams; interpret a range of mathematical information; work with numbers of any size; make calculations and check results for accuracy.

What the book contains and how to use it

This book contains 12 multiple-choice question papers, closely designed to resemble End Assessment question papers. Each paper contains 40 multiple-choice questions that test those aspects of the Standards described in the previous paragraph.

The distribution of skills tested by each of the 480 questions is tabled on *pages 207 to 210*. By including this analysis I have enabled users to easily a) select specific aspects to practise, and/or b) analyse areas of weakness from the responses to the questions. This is, of course, useful when determining which aspect needs to be strengthened before any multiple-choice paper is attempted.

It is expected that candidates, when ready to attempt a full paper, will work through each sample paper following the QCA guidelines of 1 hour and fifteen minutes and that conditions similar to the "live" End Assessment experience will prevail. Namely, the candidate is not allowed to use a calculator; the candidate works alone, and; the candidate completes an Answer Sheet, an example of which is included on *page 215*. I am aware that some candidates will do an on-line test and thus they will have no involvement with the included style of Answer Sheet. However, for those who take a paper-based test, the style of the Answer Sheet included in this book will prepare them when they do the written test paper and I believe it is the easiest way for candidates to select their answers and for tutors to mark their responses.

The answers to each question, on each paper, are tabled on *pages 211 to 213*.

The "pass" mark

For Key Skills and Adult Numeracy, QCA guidelines suggest a pass mark of between 26 to 28 out of 40. This represents a pass percentage of 65 to 70. In practice, I always determine a "Pass" is represented by 29 to 30 out of 40, which is 72 to 75%. I do this because it allows candidates some "slippage" in the real test paper. Along with thorough coverage and practice of the Standards related to the test paper, it is an assessment criterion which I have never regretted adopting.

The importance of practising for the test paper

Always make sure that your candidates have the opportunity to practise many test papers. Practice does encourage an understanding of the techniques involved — interpreting information from tables, charts and graphs; using calculations with fractions, decimals and percentages; amounts and proportion; mean, mode, median and range; metric and imperial measures and conversion; currency; time and temperature; weight and capacity; area, perimeter and volume; ratio and scale; levels of accuracy and estimation; formulae. Practice can engender confidence and competence. Competence does lead to success.

Practice Multiple-choice Paper
Suitable for:

Key Skills Level 2 Application of Number
Level 2 Adult Numeracy

Paper One

YOU NEED

- This test paper.
- A pen.
- A pencil and eraser.
- An Answer Sheet.
- A ruler marked in centimetres and millimetres

You may NOT use a calculator.
You may use a bilingual dictionary.
There are 40 questions on this paper. Try to answer ALL the questions.
When you have completed the questions you must check your answers,
then check them again.

YOU HAVE QUARTER OF AN HOUR TO READ THE PAPER
AND ONE HOUR TO COMPLETE THE 40 QUESTIONS

INSTRUCTIONS

- Make sure you write your name and today's date on the Answer Sheet.
  Use a pen to do this.
- Use a pencil to mark your answers so if you change your mind you can
  erase your choice and select another.
- Make sure that for each question you have only selected one answer.
  If you select more than one, the answer will not be marked.
- Read each question carefully before you select an answer.

Note for learners and tutors: This is a practice test that has been designed to
closely resemble the questions and question styles of a "live" paper.

Questions 1 to 5 relate to the table below, which shows the temperatures for a number of UK cities for a day in January.

1   The table below shows the temperatures for some UK cities for a day in January.

City	Temperature in degrees Centigrade (°C)
Swansea	1
London	0
Edinburgh	-7
Belfast	-5
Yeovil	4

When sorted in order of temperature, from coolest to warmest, the cities will be arranged in which order?

A   Swansea, London, Yeovil, Belfast, Edinburgh
B   London, Swansea, Yeovil, Edinburgh, Belfast
C   Swansea, Edinburgh, Belfast, London, Yeovil
D   Edinburgh, Belfast, London, Swansea, Yeovil

2   On Monday, the temperature in Edinburgh was -7°C. On Tuesday the temperature in Edinburgh was 3°C higher. What is the temperature in Edinburgh on Tuesday?

A   -10°C
B   10°C
C   4°C
D   -4°C

3   At 1am one morning the temperature in a garden in Belfast city is –5°C. By 4am the temperature has dropped by 3°C. What is the temperature at 4am?

A   –4°C
B   –8°C
C   4°C
D   8°C

Please go on to the next page

Questions 4 and 5 relate to this graph which represents the temperatures at various times of the day in a building in Belfast.

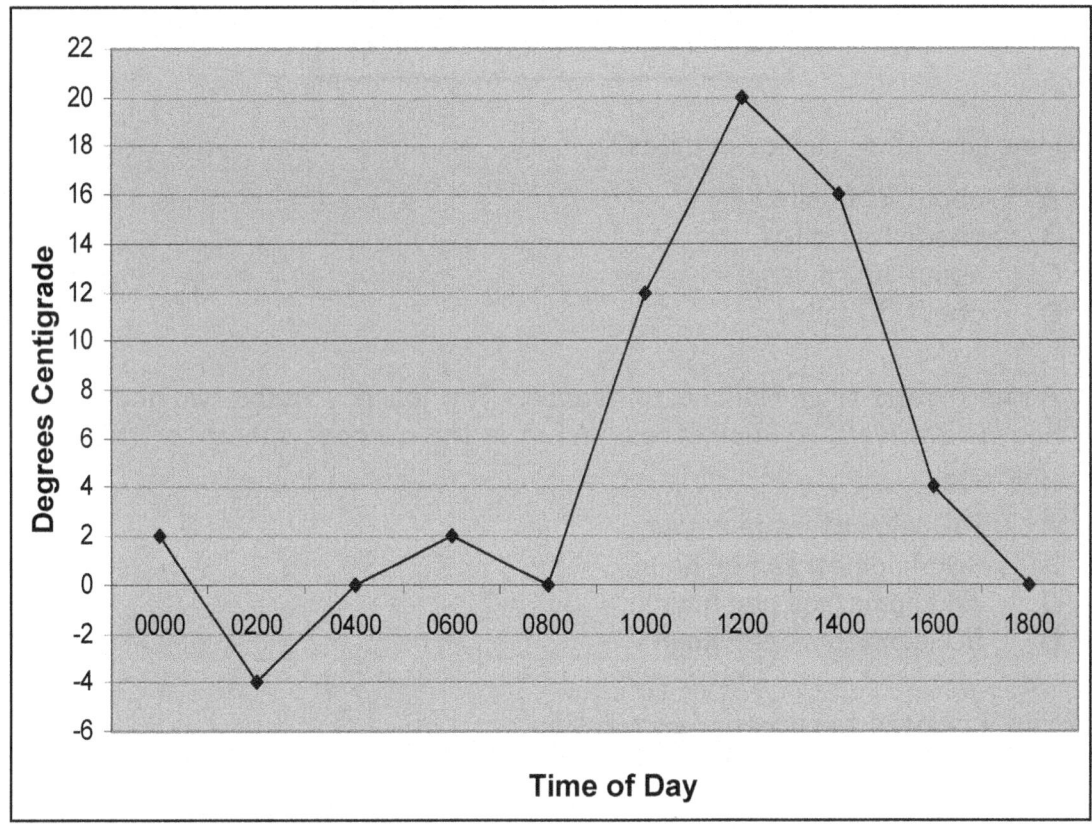

4   What is the difference between the highest and lowest temperatures?

    A   14°C
    B   18°C
    C   24°C
    D   26°C

5   According to the graph above, what is the best estimate of the temperature at 3pm?

    A   -2°C
    B   -2.5°C
    C   8°C
    D   10°C

Questions 6 to 8 relate to mileage records.

6   A businesswoman records her total mileage in one year, as shown:

    Business mileage: 6,800
    Private mileage: 10,200

    What is the ratio of her business mileage to her private mileage?

    A   3 : 4
    B   4 : 3
    C   2 : 3
    D   3 : 2

7 A businessman travels from Holland to France. The journey takes two-and a-half-hours to travel 160 kilometres.

┌─────────────────────────────────────┐
│ 8 kilometres is about 5 miles │
└─────────────────────────────────────┘

How far is his journey in miles?

A about 256 miles
B about 160 miles
C about 100 miles
D about 80 miles

8 A bus travels from Holland to France. The journey takes two and a half hours to travel 160 kilometres. What is the average speed for this journey?

A 160 kilometres per hour
B 400 kilometres per hour
C 64 kilometres per hour
D 50 kilometres per hour

Question 9 relates to currency conversion.

9 A woman from the UK spends 41 Euros in a shop. At the time, the currency conversion is approximately 1.3 Euros to the £1. To the nearest pound, the amount she spends is approximately ...?

A £13
B £30
C £32
D £58

Questions 10 to 13 relate business management information.

10 A total of 2400 people work for a company. The ratio of men to women employees is 3:5. How many **more** women than men are there in the company?

A 600
B 400
C 900
D 1500

11 A small business employs 3 men and 4 women. The men's ages are 19, 37 and 46. The women's ages are 20, 27, 27 and 50.

Which of the following statements is correct?

A the mean age of the men is less than the mean age of the women
B the mean age of the men is greater than the mean age of the women
C the median age of the men is less than the median age of the women
D you can't compare because there are different numbers of women and men

12   In a small company: the range of women's salaries is £30,000 - the range of the men's salaries is £20,000.

Which one of the following statements is correct?

A   the men earn more than the women
B   the spread of men's salaries is less than women's
C   the women earn more than the men
D   the men's salaries are more varied than women's

13   A company pays overtime at a rate of "time and a quarter". This means:

Hourly Overtime rate = 1¼ x (Normal rate per hour)

The normal payment rate is £5.80 per hour. One employee works his normal 40-hour week plus an extra 4 hours overtime.

How much is he paid for the week?

A   £232.00
B   £255.20
C   £261.00
D   £278.40

Questions 14 to 16 relate to interpreting information from tables and charts.

14   The number of hours worked by employees of a company in one week is shown in the table:

Hours worked (to the nearest hour)	Number of employees
1-10	22
11-20	12
21-30	46
31-40	144
41-50	16

Employees who work 30 hours or fewer are classed as part-time employees. What fraction, of the 240 employees, is part-time?

A   17/120
B   1/4
C   1/30
D   1/3

Please go on to the next page

15 The table shows the number of hours worked by a company's employees in one week.

Hours worked (to the nearest hour)	Number of employees
1-10	22
11-20	12
21-30	46
31-40	144
41-50	16

The pie chart represents the information from the table:

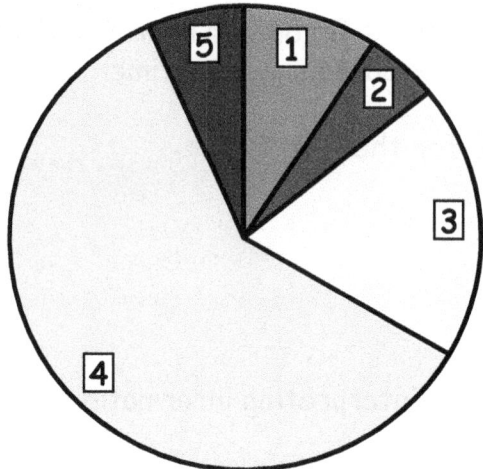

Which section represents the group working 1-10 hours per week?

A   4
B   3
C   2
D   1

16 The performance of four companies, over a two-year period, is compared in the following table:

Company	2005	2006
Westbrook	Loss of £3.0 million	Loss of £5.5 million
Newton-Smith	Loss of £2.5 million	Profit of £2 million
Hudson's	Profit of £4.5 million	Profit of £1 million
Vitaforce	Loss of £0.5 million	Profit of £0.5 million

Which company showed the greatest increase in profit between the years 2005 and 2006?

A   Westbrook
B   Newton-Smith
C   Hudson's
D   Vitaforce

Question 17 relates to unit costing of goods for sale.

17   A formula is used to work out the percentage profit made from items that are sold:

$$\frac{(£S - £C)}{£C} \times 100$$

£C is the cost price and £S is the selling price of an item for sale. If the cost price of a jacket is £20 and the selling price is £30, what is the percentage profit?

A   30%
B   33.3%
C   40%
D   50%

18   Points on this diagram show the selling price and the cost price of items for sale in a department store.

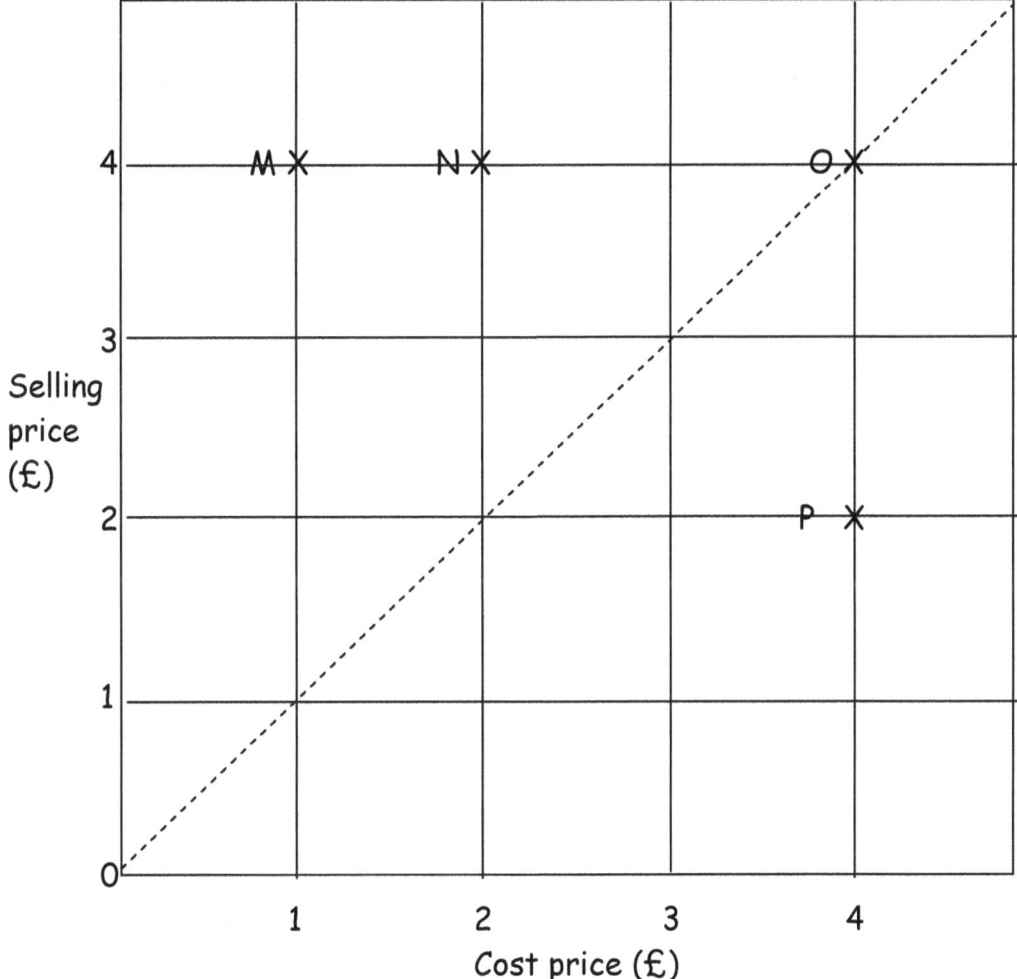

Which of these items makes the most profit?

A   M
B   N
C   O
D   P

Questions 19 to 21 relate to the work in a florist's shop.

19   A florist buys 175 bunches of carnations. Each bunch has 30 flowers.
     Which is the nearest estimate of the total number of flowers bought?

     A   200
     B   520
     C   2,600
     D   5,200

20   A florist buys 175 bunches of carnations at 50p a bunch. She sells 170
     bunches at £1.50 bunch. 5 bunches remain unsold.

     Which of the following calculations can be used to find the profit from the
     sale of the flowers in £s?

     A   170 x 1.50 - 175 x 50
     B   175 x 0.50 - 170 x 1.50
     C   170 x 1. 50 - 175 x 0.50
     D   175 x 0.50 - 175 x 1.50

21   The florist feeds her stock of cut flowers with 8 drops of fertiliser to
     every 500ml of water. How many drops of fertiliser does she need for 2.5
     litres of water?

     A   25
     B   28
     C   32
     D   40

Please go on to the next page

Questions 22 to 25 relate to the construction of a new community centre.

22 The plans for a community centre are drawn in the ratio of 1 : 50. On the drawing, the top of the roof is 16cm above the ground. How high is the roof from the ground when the community centre is built?

    A   80cm

    B   8m

    C   16m

    D   80m

The new community centre includes the design of an ornamental water feature. The outline of an L-shaped pond is shown here:

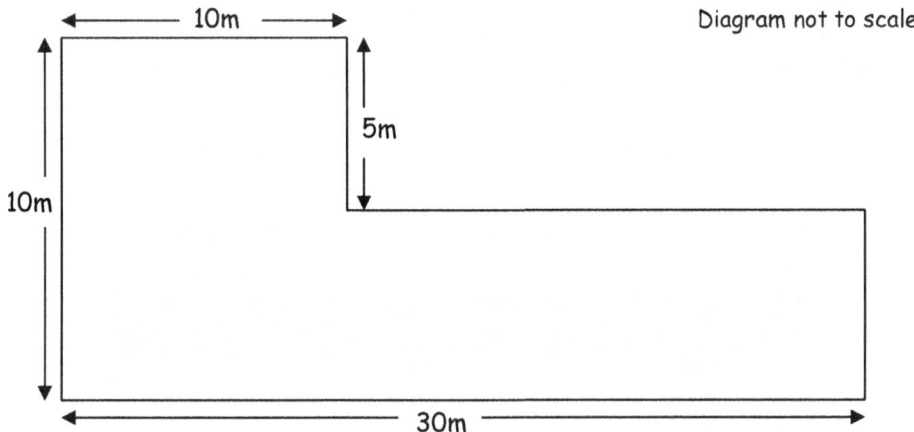

Diagram not to scale

A low fence runs all the way round the edge of the pond.

23 What is the total length of the perimeter fence?

    A   55m

    B   60m

    C   75m

    D   80m

24 What is the surface area of the water in the pond?

    A   80m²

    B   100m²

    C   200m²

    D   300m²

25 The pond holds 150 litres of water. The water is pumped in, at a rate of 5 litres a minute, through a hose from a domestic water tap. How long does it take to fill the volume of the pond?

    A $\frac{1}{4}$ of an hour

    B $\frac{1}{3}$ of an hour

    C $\frac{1}{2}$ of an hour

    D $\frac{3}{4}$ of an hour

Question 26 involves digital measurement.

26   The diagram shows the reading on a digital distance measurer.

What is the distance correct to the nearest 10 metres?

A   1.3km
B   1.36km
C   1.37km
D   1.4km

Questions 27 to 29 are about the weights of babies and adults.

27   The table shows the weight in kilograms of 45 new born babies.

Birth Weight (Kg)		
1.86	3.29	3.59
2.13	3.31	3.62
2.17	3.42	3.63
2.20	3.48	3.65
2.53	3.48	3.66
2.65	3.50	3.69
2.83	3.51	3.71
2.83	3.52	3.76
2.90	3.52	3.77
3.01	3.53	3.78
3.12	3.54	3.81
3.15	3.54	3.84
3.17	3.54	3.91
3.24	3.55	4.03
3.29	3.56	4.15

The range of the babies' birth weights is ...?

A   1.43kg
B   1.86kg
C   2.29kg
D   3.39kg

28 Referring the table of birth weights, what is the ratio of the number of babies weighing less than three and a half kilograms to the number of babies weighing three and a half kilograms and over?

A   8 : 7
B   7 : 8
C   4 : 5
D   5 : 4

29 The table shows recommended body weight for adults of average build.

Recommended Weigh for Adults of Medium Build

	Height (without shoes)			Weight (without clothes)	
	ft	ins	cms	lb	kg
Women	5	0	152	102 – 114	46 – 52
	5	1	155	105 – 117	48 – 53
	5	2	157	109 – 121	49 – 55
	5	3	160	111 – 125	50 – 57
	5	4	163	115 – 130	52 – 59
	5	5	165	119 – 134	54 – 61
	5	6	168	123 – 138	56 – 63
	5	7	170	127 – 142	58 – 65
	5	8	173	131 – 146	60 – 66
	5	9	175	135 – 146	61 – 68
	5	10	178	139 – 154	63 – 70
Men	5	5	165	122 – 135	55 – 61
	5	6	168	126 – 139	57 – 63
	5	7	170	130 – 144	59 – 65
	5	8	173	134 – 148	61 – 67
	5	9	175	138 – 152	63 – 69
	5	10	178	142 – 157	65 – 71
	5	11	180	146 – 162	66 – 74
	6	0	183	150 – 167	68 – 76
	6	1	185	154 – 172	70 – 78
	6	2	188	159 – 177	72 – 80
	6	3	191	164 – 182	75 – 83

What is the recommended weight in kilograms for a man who is 178 centimetres tall?

A   139-154
B   142-157
C   60-66
D   65-71

The following table shows sections of the population in percentages by age and gender:

Population By Age and Gender								
Percentage	Under 16	16-24	25-34	35-44	45-54	55-64	65-74	75 and above
Female								
1901	31	20	16	12	9	6	4	2
1931	23	17	16	14	12	9	6	2
1951	22	13	12	13	14	12	9	5
1991	19	12	15	13	11	10	9	9
1999	20	10	15	14	13	10	9	9
Males								
1901	34	20	16	12	9	6	3	1
1931	26	18	16	13	12	9	5	2
1951	25	14	13	14	14	11	6	3
1991	21	14	16	14	12	10	8	5
1999	21	11	15	15	13	11	8	8

30 Using information in the table, **approximately** what proportion of the male population in 1999 was aged 45 or above?

A one in twenty
B one in ten
C one in five
D two in five

31 Reading from the table above, the proportion of males aged 55 or over…

A has decreased from 1901 to 1999
B is less than the proportion of females aged 55 or over for each of the five years shown in the table
C is about six times greater in 1999 than in 1901
D is greater than the proportion of females aged 55 or over for only four of the five years shown in the table

Questions 32 to 34 relate to the breakfast cereal muesli and involve estimation.

An amount of muesli is to be weighed on these scales.

Diagram not to scale

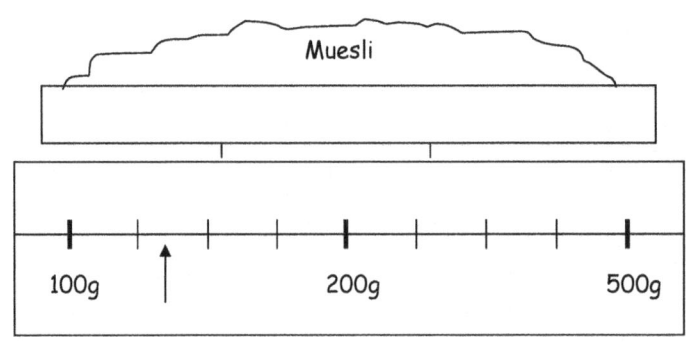

32   Which answer is closest to the weight of muesli on the scale?

   A   112g
   B   124g
   C   140g
   D   150g

33   The diagram shows a pack of muesli.

                                             Diagram not to scale

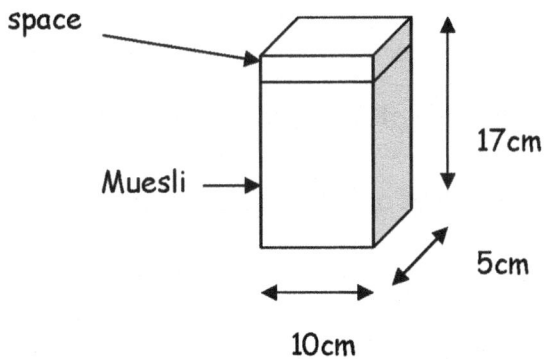

space

Muesli

17cm

5cm

10cm

100cm³ of muesli weighs 50g. The pack is filled leaving an air space 2cm deep at the top of the packet.

What is the weight of muesli in the pack?

   A   375g
   B   425g
   C   750g
   D   850g

34   A pack contains 450g of muesli. A normal serving is 30g. How many servings does the pack contain?

   A   15
   B   30
   C   105
   D   1500

Questions 35 to 38 relate to activity in a food technology laboratory.

35   A recipe for the baking of a birthday cake requires a 7 inch diameter tin.

   1 inch is approximately 2.5 centimetres.

   What is the diameter of the tin in centimetres, rounded to the nearest centimetre?

   A   3.5cm
   B   18cm
   C   25cm
   D   50cm

36 The baking instructions for a cake state: "Pre-heat the oven to 320°F". A formula converts from degrees Fahrenheit to degrees Celsius.

$$°C = (°F - 32) \times 5/9$$

Using the formula, 320°F is ...?

A   140°C
B   150°C
C   160°C
D   170°C

Question 37 and 38 relates to interpreting a diagram and using a formula.

37 The diagram shows a block of cooking chocolate.

The formula to find the volume of the block is:

Volume = length x width x height, or, v = l x w x h

Diagram not to scale

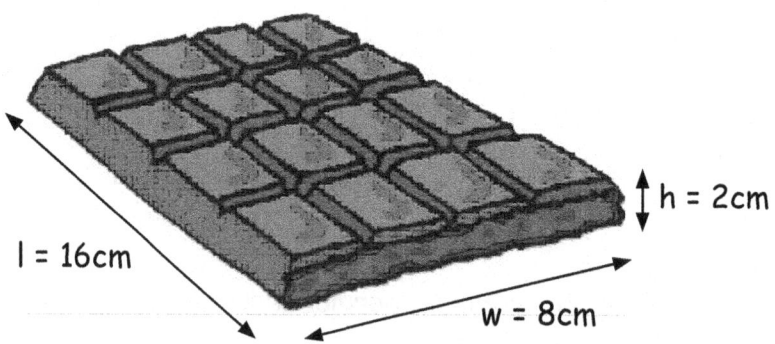

l = 16cm

w = 8cm

h = 2cm

Using the formula, the volume of the chocolate block is:

A   256cm³
B   456cm³
C   600cm³
D   800cm³

38 A series of cookery books is priced at £6.99 each. The bill for 12 books comes to £75.49 and includes a discount of 10% on the total. Which is the most appropriate method to check the calculation?

A $\dfrac{75.49}{12} \times 1.1$

B $\dfrac{75.49}{12} \times 0.9$

C $(6.99 \times 12) \times 1.1$

D $(6.99 \times 12) \times 0.9$

Questions 39 and 40 relate to information about leisure pursuits.

39 In the game of golf, the net score of each player can be worked out by subtracting their handicap from their gross score. In a golf tournament, the scores were recorded as follows:

Player	Handicap	Gross Score
Alison	14	90
Jean	14	84
Nigel	13	89
Ranjid	12	85

Which of these players has the lowest net score?

A   Alison
B   Jean
C   Nigel
D   Ranjid

40 A sign in the art gallery window shows the opening times. How many hours per week (Monday – Sunday) is the museum open between April and October?

LONG MEGTON GALLERY

Opening Times April – October

Monday – Friday:	10am – 12 noon, 1pm – 5pm
Saturdays and Sundays:	9am – 12 noon, 1pm – 4pm

A   40
B   42
C   47
D   49

End of Paper

Practice Multiple-choice Paper
Suitable for:

Key Skills Level 2 Application of Number
Level 2 Adult Numeracy

Paper Two

YOU NEED

- This test paper.
- A pen.
- A pencil and eraser.
- An Answer Sheet.
- A ruler marked in centimetres and millimetres

You may NOT use a calculator.
You may use a bilingual dictionary.
There are 40 questions on this paper. Try to answer ALL the questions.
When you have completed the questions you must check your answers, then check them again.

YOU HAVE QUARTER OF AN HOUR TO READ THE PAPER
AND ONE HOUR TO COMPLETE THE 40 QUESTIONS

INSTRUCTIONS

- Make sure you write your name and today's date on the Answer Sheet. Use a pen to do this.
- Use a pencil to mark your answers so if you change your mind you can erase your choice and select another.
- Make sure that for each question you have only selected one answer. If you select more than one, the answer will not be marked.
- Read each question carefully before you select an answer.

Note for learners and tutors: This is a practice test that has been designed to closely resemble the questions and question styles of a "live" paper.

Questions 1 to 3 are about temperature records in a commercial greenhouse
A market gardener records daily temperatures in °C.

The table shows temperatures recorded during four weeks in January.

Week beginning:	Mon	Tues	Wed	Thurs	Fri	Sat	Sun
01 January 2007	1	-1	-3	-4	-1	0	3
08 January 2007	6	5	6	3	2	-1	0
15 January 2007	3	-2	0	0	4	7	7
22 January 2007	4	2	5	5	2	4	5

1   When the temperature in the greenhouse drops to 0°C a thermostat
     control switches on a heater. In the four weeks in January, for what
     fraction of days was the heater switched on?

   A $\dfrac{1}{8}$

   B $\dfrac{5}{14}$

   C $\dfrac{7}{31}$

   D $\dfrac{2}{7}$

2   The range of temperatures in the four weeks in January was ...?

     A   11°C
     B   8°C
     C   4°C
     D   -4°C

3   What percentage of days in the four weeks in January was the
     temperature 3°C or higher?

     A   7%
     B   25%
     C   50%
     D   75%

Please go on to the next page

Questions 4 and 5 are about a company's salaries.

A Human Resources Assistant collects data to compare salaries of men and women who work part-time hours in the company.

Women's Salaries (£ per month)	Men's Salaries (£ per month)
680, 680, 680, 760, 760, 840, 920, 920, 1240, 1360	680, 720, 760, 760, 800, 840, 920, 920, 960, 1040, 1840, 2000

4   What is the range of men's salaries?

    A   680

    B   1040

    C   1320

    D   1840

5   The Assistant works out that the mean salary for women employees is £884. How much more than this is the mean salary for men?

    A   126

    B   136

    C   148

    D   160

Questions 6 and 7 are about a motoring survey.

6   In a survey of 800 drivers using a city centre car park, 300 drivers parked in the car park at least four times a week. The percentage of drivers using this car park at least four times per week is …?

    A   23.5%

    B   25.0%

    C   33.3%

    D   37.5%

7   On the first day of the survey a total of 252 car drivers used the car park. From 7am to 9am, 147 cars came into the car park. What is the nearest fraction of cars that parked between 7am and 9am?

    A $\dfrac{3}{8}$

    B $\dfrac{2}{5}$

    C $\dfrac{3}{5}$

    D $\dfrac{5}{8}$

Questions 8 to 10 are about a mobile hairdressing and beauty therapy business.

The owner of a mobile hairdressing and beauty therapy business needs to have some leaflets printed to advertise her business.

The printer's price list shows the costs as follows:

Printing Service	Cost
Initial set-up charge, per design	£20
First 200 leaflets	£5 per 100 leaflets
Additional leaflets over 200	£2.50 per 100 leaflets

8   Which of the following formulae shows the cost of printing 500 leaflets?

A   Cost (£s) = 20 + (5 × 10)
B   Cost (£s) = 20 + (2 × 5) + (3 × 2.5)
C   Cost (£s) = 20 + (2.5 × 10)
D   Cost (£s) = 20 + (7.5 × 10)

9   The hairdresser has a budget of £60 for printing costs. She requires two different leaflet designs to be printed: one for the hairdressing prices and one for the beauty therapy prices. How many leaflets **in total** can she have printed for £60?

A   200
B   400
C   600
D   800

10   A local newspaper sells advertising space at a discount rate for new businesses. The hairdresser chooses a half column advertisement that normally costs £42 and the discount rate is £14, making a saving of £28. Which calculation shows the amount **saved** as a percentage of the original rate?

A $\dfrac{42}{14}$ × 100%

B $\dfrac{42}{28}$ × 100%

C $\dfrac{14}{42}$ × 100%

D $\dfrac{28}{42}$ × 100%

Please go on to the next page

Questions 11 to 15 are about a school's table tennis tournament.

The floor plan of the sports hall booked for the tournament is shown below.

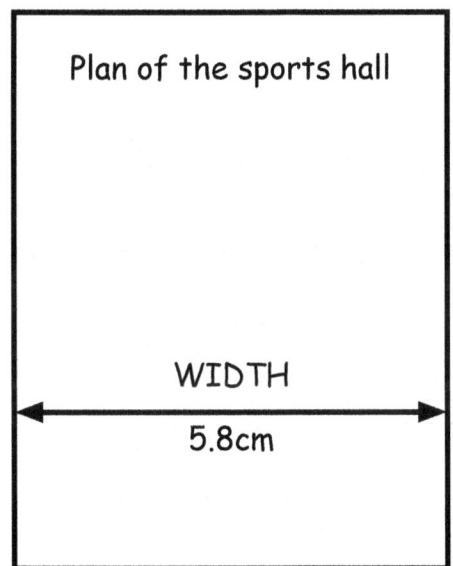

Scale: 10mm equals 4m

11 Calculating from the scale drawing, what is the width of the full-sized sports hall?

A   2.3m
B   4.6m
C   23.2m
D   46.4m

12 A refreshments table is to be set up at the back of the sports hall. The table measures 1.4m by 0.7m. On the plan, the size of the table, drawn to scale, should be ...?

A   2.8mm by 1.4mm
B   3.5mm by 1.75mm
C   35mm by 17.50mm
D   28mm by 14mm

13 The cost of printing tickets for the tournament includes the setting-up fee of £9.75 plus £1.75 for every 100 tickets printed. Which one of the following calculations shows the best estimate of the total printing costs for 300 tickets?

A   2 + 3 + 10
B   10 + (2 × 3)
C   (10 + 3) × 2
D   2 × 3 × 10

14 High-energy fruit squash is sold between matches. The drinks are made up of 1 part fruit squash concentrate to 3 parts water.

How much of the concentrate is needed to make 2 litres of fruit squash?

A   50ml
B   60ml
C   500ml
D   600ml

15 A summary of the total costs and income from the tournament is shown below:

Tournament preparation costs	£
Printing programmes, tickets and posters	20
Hire of sports hall	75
Hire of tables and table tennis equipment	40
Refreshments	9
Tournament income from sales	£
Tickets	164
Refreshments	22

Which of the following calculations shows the profit made from the tournament event?

A   £(20 + 75 + 40 + 9 - 164 - 22)
B   £(20 + 75 + 40 + 9) - (164 - 22)
C   £(164 + 22 - 20 + 75 + 40 + 9)
D   £(164 + 22) - (20 + 75 + 40 + 9)

Questions 16 to 18 are about a sponsored velodrome cycle race.

A total of 225 cyclists applied to participate in a sponsored cycle race to raise funds for a national youth campaign.

To take part in the race event each participant had to show they had at least £100 of sponsorship. One month before the event, 75 potential participants had not yet reached £100 in sponsorship.

16 What was the ratio of those with £100 sponsorship to those without sponsorship?

A   1 : 2
B   3 : 1
C   2 : 1
D   5 : 1

The diagram shows the cycle circuit in the velodrome where the sponsored race took place.

Diagram not to scale

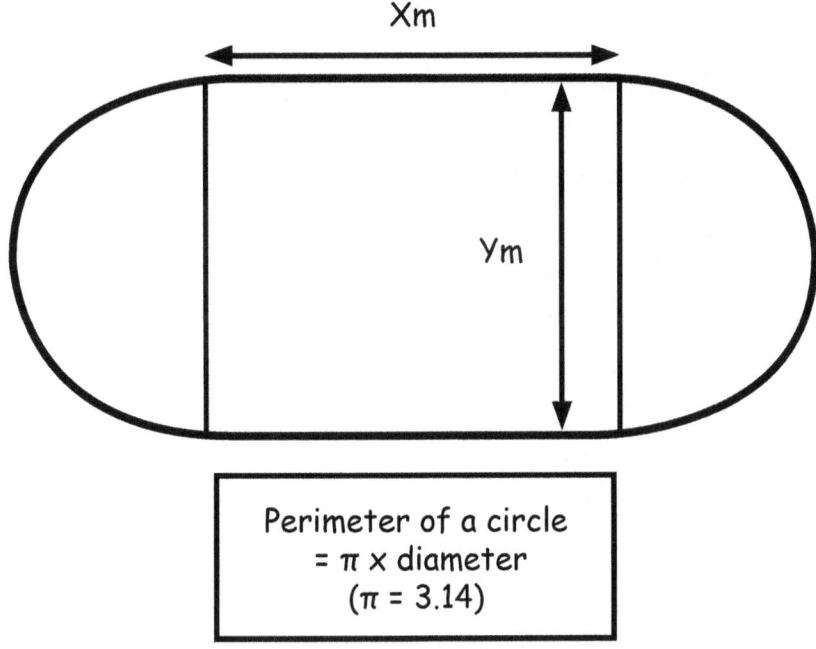

Perimeter of a circle
= π x diameter
(π = 3.14)

17 Which one of the calculations works out the approximate total perimeter length of the cycle circuit, in metres?

A (2 x Xm) + 2 x (π x Ym)
B (Xm + Ym) x (2 x π)
C (2 x Ym) x 2 x (π x Xm)
D (Xm x Ym) + (2 x π)

18 200 cyclists took part in the sponsored race. Their finishing times varied and the mode of the finishing times was 2 hours exactly.

Of the following statements, which **must** be true?

A everyone takes 2 hours
B the total time taken is 400 hours
C the most common finishing time was 2 hours
D half of the entrants take less than 2 hours

Questions 19 and 20 relate to the amount of diesel in the tank of a builder's truck.

The fuel tank of a delivery truck holds 24 gallons.

19 How many litres of diesel does the delivery truck's tank hold, given that one gallon is equivalent to 4.55 litres?

A 109.2 litres
B 104.9 litres
C 6.9 litres
D 5.3 litres

The diagram below shows the truck's fuel gauge reading.

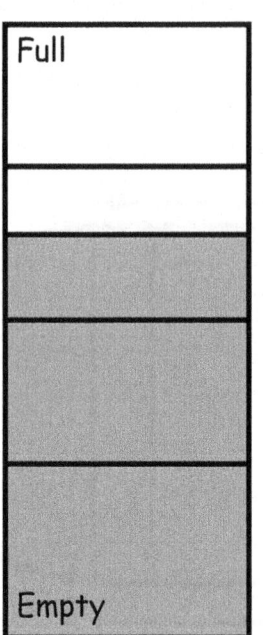

Diagram not to scale

20   To the nearest gallon, how much diesel is left in the tank?

  A   14 gallons
  B   16 gallons
  C   18 gallons
  D   20 gallons

21   A mother wishes to weigh her two-year-old son, who will not stand still on the digital weighing scales in the bathroom.

First, she weighs herself:

Then, she weighs herself while holding the two-year-old.

How much does her son weigh?

  A   5.56kg
  B   5.57kg
  C   5.58kg
  D   6.52kg

Questions 22 to 25 are about the design of a care home garden.

22 The garden designer draws a plan to work from, using a scale of 1 : 40. The length of the garden is 22 metres.

What is the length of the garden in the scale drawing?

A 88.0cm
B 55.0cm
C 8.8cm
D 5.5cm

23 A marble sculpture, weighing 35 stones, is to feature in the garden. 1 stone is approximately equal to 6.4 kilograms. What is the weight of the marble sculpture to the nearest kilogram?

A 207kg
B 224kg
C 237kg
D 244kg

The garden designer includes a circular pond for goldfish. The diagram below shows a plan view of the pond.

diagram not to scale

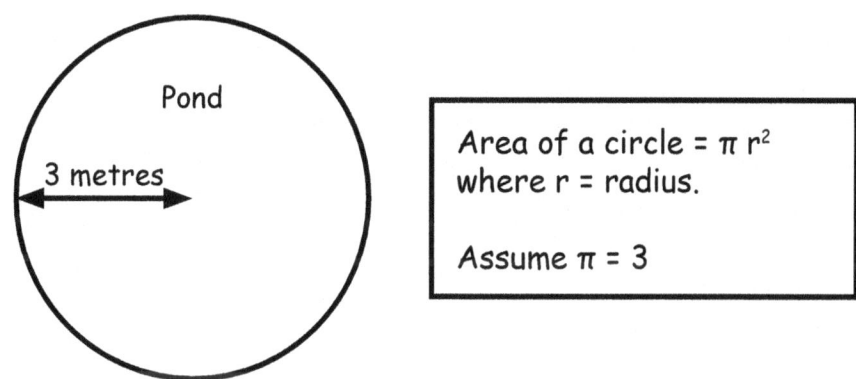

Pond

3 metres

Area of a circle = πr^2
where r = radius.

Assume $\pi = 3$

The pond must be covered with wire mesh to comply with health and safety regulations.

24 What is the minimum area of wire mesh required to cover the surface of the pond?

A 144m²
B 72m²
C 27m²
D 18m²

25 The total area of the garden is 220m². 132m² of the garden will be covered by grass lawn. What fraction of the garden area will need to be sown with grass seed?

A $\dfrac{1}{4}$

B $\dfrac{2}{5}$

C $\dfrac{3}{5}$

D $\dfrac{4}{5}$

A glassmaker works out the dimensions of a paperweight to be cast in the shape of a **square-based pyramid**.

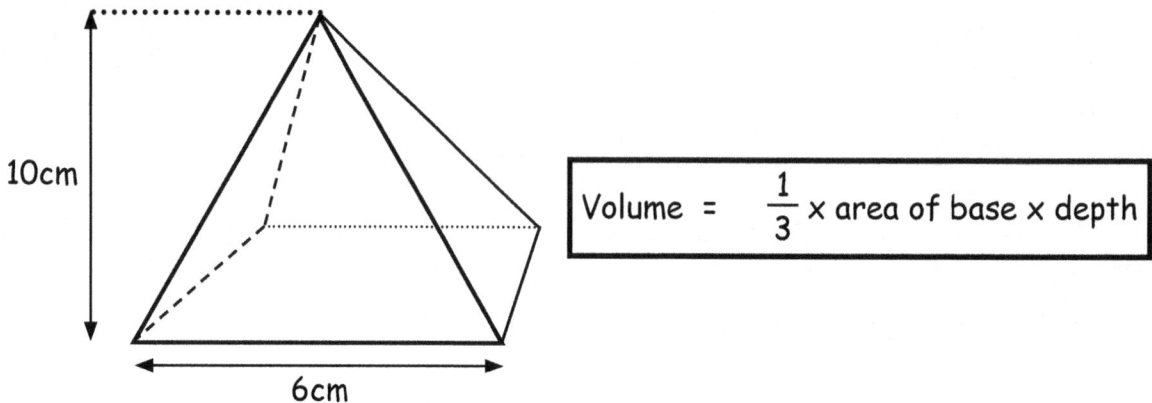

10cm

6cm

Volume = $\dfrac{1}{3}$ x area of base x depth

Diagram not to scale

26 From the dimensions given on the diagram, what is the volume of the glass paperweight?

A   360cm
B   240cm
C   160cm
D   120cm

Please go on to the next page

Questions 27 to 30 are about food preparation in a commercial food laboratory.

A portion of cake is weighed on kitchen scales.

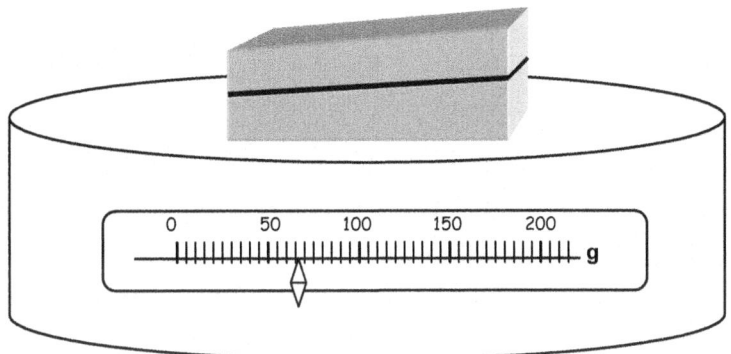

27 How much does this piece of cake weigh?

    A   55.5g
    B   67.0g
    C   70.5g
    D   75.0g

28 A food technologist measures the time it takes to heat up a vat of white sauce to 95°C for a new lasagne recipe. The temperature is taken every 10 seconds and the results recorded in a table:

	Time (seconds)	Temperature °C
Start	0	35
	10	43
	20	51
	30	59
	40	67
	50	74
	60	80
	70	84
	80	87
	90	91
	100	93
	110	94
End	120	95

During the cooking time, the temperature of the sauce rises by 60°C. The total heating time is 120 seconds.

What percentage of the total rise in temperature takes place during the first half of the cooking time?

    A   50%
    B   75%
    C   80%
    D   85%

29 A production run of plain sponge in the bakery uses 12 stones of flour. 1 stone is approximately 6.4 kilograms. How many kilograms of flour, to the nearest kilogram, are used?

A 77kg
B 73kg
C 20kg
D 19kg

30 The plain sponge recipe is modified to make chocolate sponge by replacing some of the flour with cocoa powder. $10\frac{1}{2}$ stones of flour are used. What fraction of the original amount of flour is replaced with cocoa powder?

A $\frac{1}{8}$

B $\frac{3}{8}$

C $\frac{5}{8}$

D $\frac{7}{8}$

Please go on to the next page

Questions 31 and 32 are about a diet plan.

James starts a new weight-reduction diet where his maximum daily allowance is 1900 calories.

Calorie values for a range of foods are shown in the following table:

Amount	Food item	Calories
100ml	Orange Juice	45
half	Grapefruit	30
100ml	Milk	49
per slice	Toast	60
25g	Marmalade	64
10g	Butter	74
25g	Bran Flakes	86
1 medium size	Boiled Egg	92
any	Tea (without milk or sugar)	0
5ml teaspoon	Sugar	18

For breakfast on day 1 of the diet, James can have half a grapefruit, 50 grams of bran flakes with 150ml milk, and a cup of tea with 2 teaspoons of sugar.

James needs to know the number of calories taken at breakfast as a percentage of his total daily calories.

31   Which calculation should he use to work it out?

A $$\frac{(86 \times 2) + (49 \times 1.5) + (18 \times 2) + (30)}{1900} \times 100$$

B $$\frac{1900}{(86 \times 2) + (49 \times 1.5) + (18 \times 2) + (30)} \times 100$$

C $$\frac{86 + 49 + 18 + 30}{1900} \times 100$$

D $$\frac{1900}{86 + 49 + 18 + 30} \times 100$$

32   Which combination of the breakfast foods in the table has the most calories?

A   1 x half grapefruit, 1 slice of toast and 1 boiled egg
B   100ml orange juice, 50 grams of bran flakes and 100ml milk
C   2 slices of toast, 10g butter and 25g marmalade
D   1 x half grapefruit, 1 cup of tea with no milk and 1 teaspoon of sugar

33 A school photographer designs a presentation frame from stiff card to display individual photographs of pupils. The measurements he uses are shown in the diagram.

Presentation frame

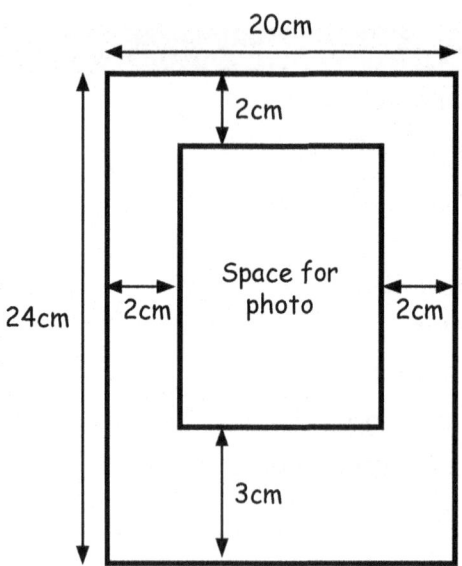

Diagram not to scale

To check the area to display the photograph (in cm²), which of the following calculations should he use?

A   (24 – 5) x (20 – 2)
B   (24 – 3) x (20 – 4)
C   (24 – 2) x (20 – 2)
D   (24 – 5) x (20 – 4)

34 Sonia wants to paint three walls of her room with pink matt emulsion. Each wall measures 2.5m x 3.5m. A 2.5 litre tin of paint covers 25 square metres and for best results, two coats of paint are required.

What is the **minimum** number of tins of paint Sonia needs to buy?

A   1 tin
B   2 tins
C   3 tins
D   4 tins

35 At a county fair, in a competition to guess the amount of milk contained in a sealed, metal milk churn, five farmers give their estimates:

$$0.85, \quad 67\%, \quad \frac{3}{4}, \quad \frac{4}{5}, \quad 0.7$$

Which estimate has the smallest value?

A   0.7

B $\frac{3}{4}$

C   67%

D $\frac{4}{5}$

36 In a dentist's waiting room a tank, for tropical fish, is to be installed. The shape of the fish tank is cuboid and the measurements are shown in the diagram.

Diagram not to scale

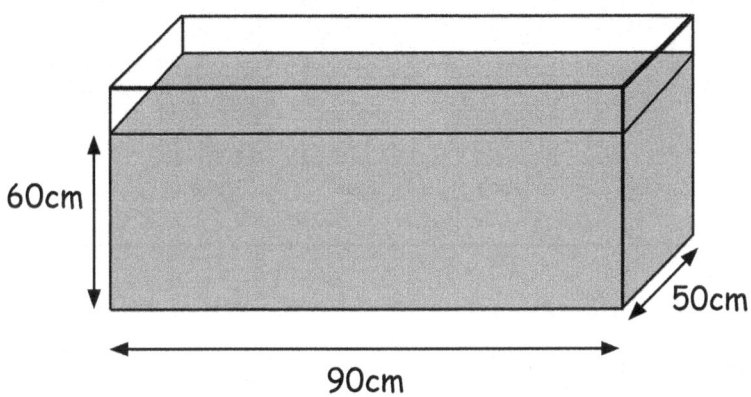

A 10-litre bucket is used to fill the tank to a depth of 60cm.

1 cubic centimetre = 1 millilitre

How many times will the 10-litre bucket have to be filled and emptied into the tank for the water to reach the 60cm mark?

A 3
B 27
C 54
D 81

Questions 37 to 40 are about a steel mill warehouse.

37 Peter has been learning how to drive a forklift truck in an industrial warehouse. The heaviest weight he has transported is 25 hundredweights.

1 ton = 20 hundredweights; 1 ton = 1020 kilograms

What is 25 hundredweights in kilograms?

A 1000kg
B 1025kg
C 1275kg
D 2500kg

38 It takes Peter approximately 4 minutes to move 1020 kilograms from the loading bay of the warehouse to the steel mill workshop and back to the loading bay. If he maintains the same speed, how long does it take Peter to move 25500 kilograms?

A 1 hour 2 minutes and 30 seconds
B 1 hour 29 minutes
C 1 hour 40 minutes
D 1 hour 55 minutes

The bar chart shows the output of steel in millions of tons over six years.

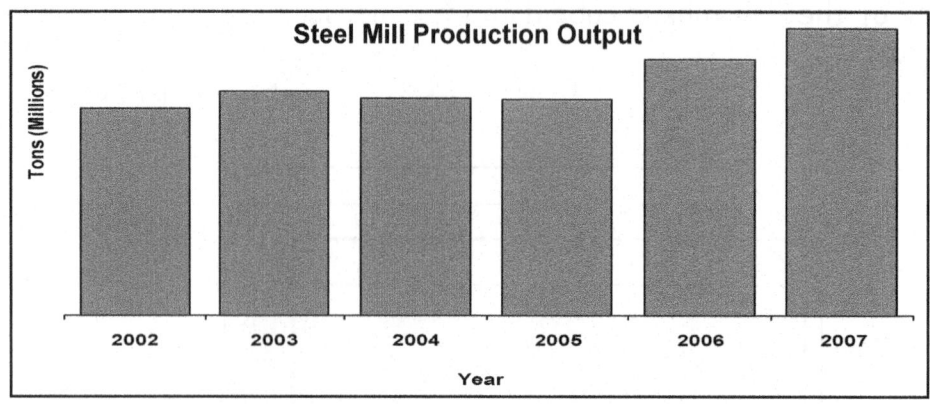

39   How is the chart misleading?

    A   the bars are the wrong width
    B   the vertical axis has no scale
    C   the bars do not increase evenly
    D   the horizontal axis shows no values

40   The chart shows which industry sectors used the steel mill's output in 2007. The total output, from the steel mill, was about 800,000,000 tons.

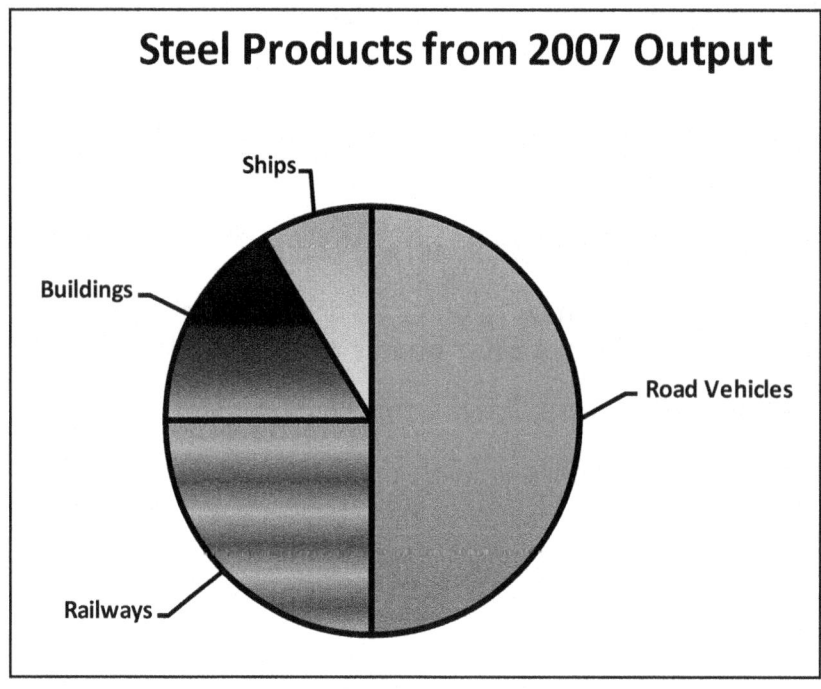

Approximately how many millions of tons of steel went into the construction of buildings from the steel mill's output?

    A   133
    B   200
    C   400
    D   500

End of Paper

Practice Multiple-choice Paper
Suitable for:

Key Skills Level 2 Application of Number
Level 2 Adult Numeracy

Paper Three

YOU NEED

- ■ This test paper.
- ■ A pen.
- ■ A pencil and eraser.
- ■ An Answer Sheet.
- ■ A ruler marked in centimetres and millimetres

You may NOT use a calculator.
You may use a bilingual dictionary.
There are 40 questions on this paper. Try to answer ALL the questions.
When you have completed the questions you must check your answers, then check them again.

YOU HAVE QUARTER OF AN HOUR TO READ THE PAPER
AND ONE HOUR TO COMPLETE THE 40 QUESTIONS

INSTRUCTIONS

- ■ Make sure you write your name and today's date on the Answer Sheet. Use a pen to do this.
- ■ Use a pencil to mark your answers so if you change your mind you can erase your choice and select another.
- ■ Make sure that for each question you have only selected one answer. If you select more than one, the answer will not be marked.
- ■ Read each question carefully before you select an answer.

Note for learners and tutors: This is a practice test that has been designed to closely resemble the questions and question styles of a "live" paper.

Questions 1 to 4 are about a local estate agent's business.

1   Out of 2500 properties sold by an estate agent in one year, 2000 are flats. What is the ratio of flats to other types of property sold?

A   1 : 4
B   4 : 1
C   2 : 3
D   5 : 4

2   Of the 2000 flats that are sold by the estate agent, 503 have only one bedroom. Approximately what percentage of one-bedroom flats are sold?

A   15%
B   25%
C   33.3%
D   45%

A new flat has come onto the market to be sold by the estate agent. The dimensions of the living room of the flat are shown in the diagram:

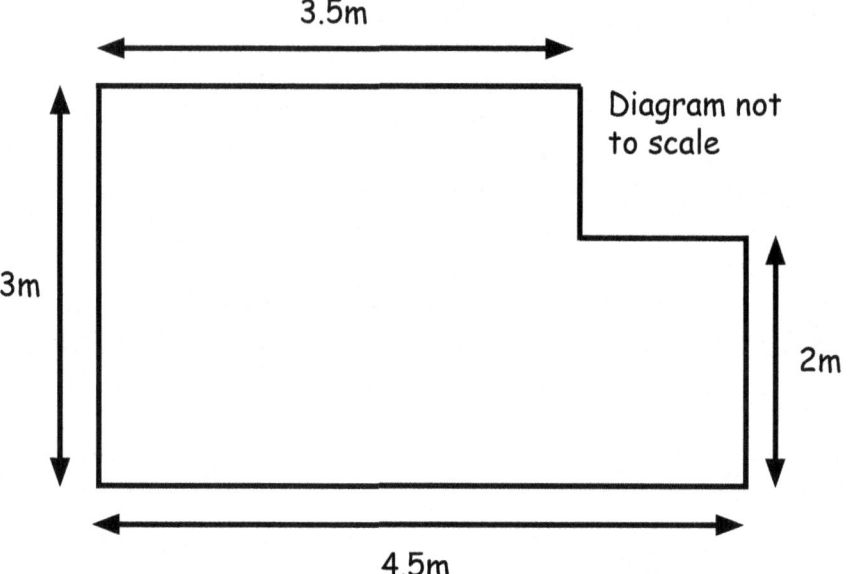

3   What is the area of the living room in square metres?

A   14.0m²
B   13.5m²
C   13.0m²
D   12.5m²

4 The estate agent has properties for sale in each of the valuation bands shown in the table:

Valuation Band	A	B	C	D	E	F	G	H
Council Tax Rates £s	725	857	979	1,102	1,347	1,592	1,837	2,225

What is the mean of the council tax rates?

A £1333
B £1600
C £2225
D £2960

Questions 5 to 9 are about an ice cream shop that is opening on the High Street.

5 The temperature in the shop is 19°C. The temperature in the freezer room is -18°C. What is the difference between the temperature in the freezer room and the temperature in the shop?

A -1°C
B 1°C
C 17°C
D 37°C

6 Leaflets, advertising the opening of the ice cream shop, cost the shop owner a fixed set-up charge of £45. Each leaflet costs 5.5 pence to print. The correct calculation to find the total cost of printing 1000 leaflets in pounds, is:

A $\dfrac{(45 + 5.5) \times 1000}{100}$

B $\dfrac{(5.5 \times 1000) \times 45}{100}$

C $\dfrac{(5.5 \times 1000) + 45}{100}$

D $(55 \times 1000) + 45$

Please go on to the next page

7 Pre-packed tubs of frozen ice cream are brought from the freezer room and stored in glass topped chiller cabinets ready for sale in the shop. The diagrams show the measurements of the one of the individual tubs and one of the chiller cabinets.

Diagrams not to scale

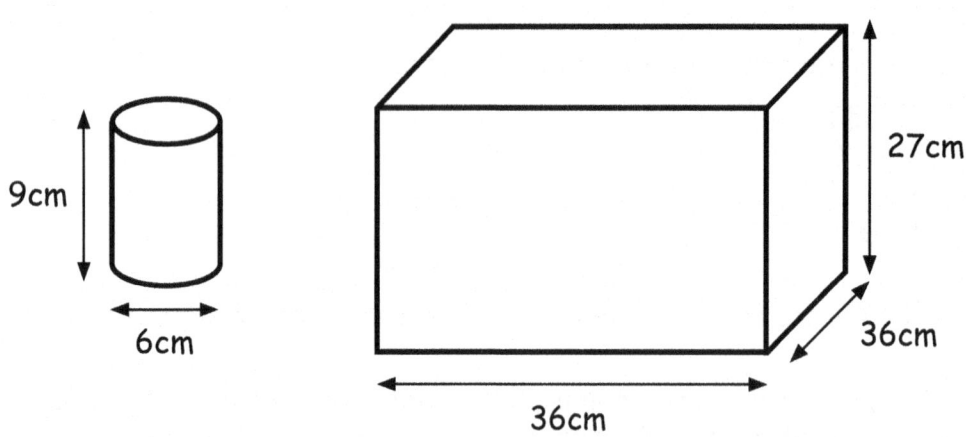

How many of tubs of ice cream will it take to fill the chiller cabinet?

A 54
B 63
C 72
D 108

8 The table shows the number of hours a shop assistant works in the shop in a week.

Day	Monday	Tuesday	Wednesday	Thursday	Friday
Hours per day	7	7	9	7	7

The hourly rate is £5.60 and a daily bonus is paid of £5 per day.

Which calculation can be used to work out the total of pay and bonus per week?

A (37 x £5.60) + (5 x £5)
B (37 x £5.60) + £5
C (£5.60 + £5) x 37
D (5 x £5.60) + (5 x £5)

9 The yoghurt-style ice cream variety contains 50 calories per ounce. There are about 28 grams in an ounce. Approximately how many calories are there in a 112g tub of the yoghurt-style ice cream?

A 150 calories
B 200 calories
C 220 calories
D 250 calories

10 The diagram shows a transparent panel in an electric kettle. The kettle holds 1.5 litres when full.

Full

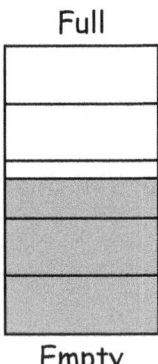

Empty

What is the nearest estimate of the amount of water in the kettle?

A   700ml
B   750ml
C   800ml
D   950ml

11 A house contents insurance policy costs £120 per year. After a claim on the policy, the cost per year has increased by 70%. How much does the contents insurance cost after the increase?

A   £84
B   £190
C   £204
D   £240

Questions 12 and 13 are about the kitchen in a nursing home.

12 A catering assistant prepares to oven-cook a large, fresh chicken for lunch in a care home. The chicken weighs 3 kilograms. Food standards authorities advise the use of a cooking time formula to be certain that meat products are properly cooked to prevent the spread of harmful bacteria.

> Chicken should be cooked in a preheated oven at 180°C for **45 minutes per kg,** plus a **further 30 minutes** in total.

Using the recommended cooking time formula, what is the cooking time of the chicken to the nearest quarter of an hour?

A   2 hours

B $2\frac{1}{4}$ hours

C $2\frac{1}{2}$ hours

D $2\frac{3}{4}$ hours

13 The cooking temperature for chicken is 180°C. To change from Celsius (°C) to Fahrenheit (°F), use the formula:

$$F = \left(\frac{9C}{5}\right) + 32$$

What is the cooking temperature in degrees Fahrenheit, to the nearest degree?

A   300°F
B   324°F
C   328°F
D   356°F

Questions 14 to 16 are about travelling and distances.

14 The distance between Lisbon and Porto is 170 miles.

> 1.6 kilometres = 1 mile

What is the distance between Lisbon and Porto to the nearest kilometre?

A   162km
B   262km
C   272km
D   282km

15 On a road map, the distance between Lisbon and Faro in the Algarve measures 5.8cm. The scale of the map is 1cm : 20km. The real distance between Lisbon and Faro is ...?

A   100.16km
B   101.6km
C   110.6km
D   116km

16 The road route from Lisbon to Porto is 170 miles. A car journey takes 3 hours and 30 minutes.

> $$\text{Average Speed} = \frac{\text{Distance}}{\text{Time}}$$

What is the average speed of the car during this journey, in miles per hour, to the nearest mile per hour?

A   29mph
B   39mph
C   49mph
D   59mph

17 The numbers of entrants in a sponsored walk event, during the Spring Bank Holiday, are recorded in the table.

Spring Bank Holiday Sponsored Walk Event			
Entrants	Sunday	Bank Holiday Monday	Total
Over 18s	70	102	172
18 and under	46	50	96
Totals	116	152	268

The entry fee for Over 18s is £3.50 and for 18-year olds and under, £1.75. What is the total income from entry fees received on Bank Holiday Monday, to the nearest pound?

A £357
B £445
C £602
D £1,407

Questions 18 to 20 are about passenger journeys on public transport in the North East of England.

18 The timetable for the Newcastle to Sunderland express bus is shown here:

Newcastle to Sunderland Monday to Friday									
Newcastle	05.39	06.12	07.11	08.23	08.47	09.26	09.56	16.56	17.26
Gateshead	05.48	06.21	07.20	08.33	08.57	09.36	–	17.06	17.40
Heworth	05.53	06.26	07.25	08.38	09.02	09.41	10.10	17.12	17.45
Felling	05.59	06.31	07.30	08.43	09.07	–	10.15	17.17	17.50
Boldon	06.04	06.37	07.36	08.49	09.13	09.49	10.20	17.23	17.55
Seaburn	06.08	06.40	07.39	08.53	09.16	09.52	10.23	17.26	17.58
Sunderland	06.12	06.45	07.44	08.57	09.20	09.57	10.28	17.31	18.03

(then these minutes past each hour until)

Sunderland to Newcastle Monday to Friday										
Sunderland	06.35	07.15	07.31	07.44	08.09	09.10	09.41	17.17	17.47	18.19
Seaburn	06.40	07.20	07.36	07.49	08.14	09.15	09.46	17.19	17.52	18.24
Boldon	06.43	07.23	07.39	07.52	08.17	09.18	09.49	17.22	17.55	18.27
Felling	06.48	–	07.43	07.57	08.22	09.23	09.54	17.27	18.00	18.32
Heworth	06.52	–	07.47	08.01	08.26	09.27	09.58	17.31	18.04	18.36
Gateshead	06.56	–	07.51	08.06	08.30	09.31	–	17.35	–	18.40
Newcastle	07.04	07.50	08.01	08.14	08.42	09.43	10.13	17.44	18.16	18.48

(then these minutes past each hour until)

Reading the timetable, how many buses leave Sunderland between 8am and noon?

A 7
B 5
C 4
D 3

19 The table shows the number of passenger journeys, over the last ten years, made on the Tyne and Wear metro system by people travelling to Sunderland over the past ten years.

Year	Passenger journeys (millions)
1997	21.31
1998	21.34
1999	21.41
2001	21.55
2002	21.71
2003	21.47
2004	21.62
2005	21.76
2006	21.95
2007	22.01

To work out the percentage increase in the number of passenger journeys between 1997 and 2007, which of the calculations will give the correct answer?

A $\dfrac{21.31}{22.01} \times 100$

B $\dfrac{22.01}{21.31} \times 100$

C $\dfrac{(22.01 - 21.31)}{21.31} \times 100$

D $\dfrac{(22.01 - 21.31)}{22.01} \times 100$

20 In a survey of metro passengers travelling to Sunderland in 2007, out of 22.01 million passengers surveyed, 7.33 million said they were satisfied with the metro service provided. What fraction of these passengers, approximately, said they were happy with the metro service?

A $\dfrac{1}{3}$

B $\dfrac{1}{5}$

C $\dfrac{1}{6}$

D $\dfrac{1}{7}$

Questions 21 to 26 are about a building company's contract to build an extension on the ground floor of a house for a disabled person's new bathroom.

21   As a guide, the builder charges £700 per square metre.

Diagram not to scale

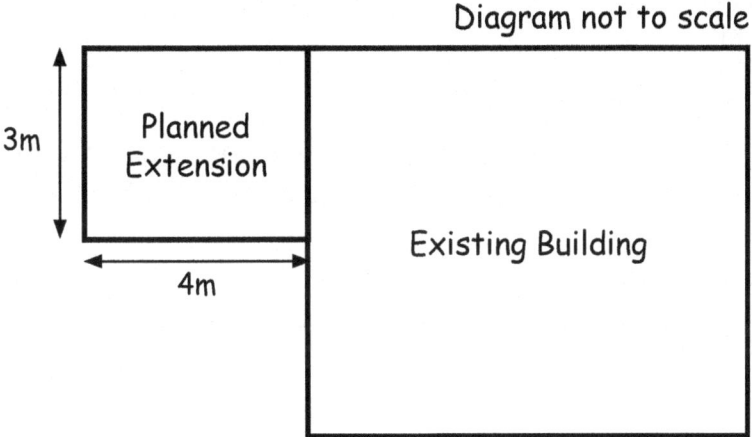

3m

Planned Extension

Existing Building

4m

What is the builder's estimate for the work?

A   £1200
B   £2800
C   £4900
D   £8400

22   The builder mixes concrete to make the floor using the proportions:

8 parts Aggregate to 3 parts Cement

The builder's concrete mixer can hold 44 shovels of materials.

To fill the mixer with the correct amounts of aggregate and cement, how many shovels of cement does he need to add?

A   8
B   12
C   22
D   33

23   The concrete is mixed using 2 litres of water to every 25 kilograms of the dry aggregate and cement materials. How much water, to the nearest litre, must be added to 175 kg of dry materials?

A   14 litres
B   13 litres
C   12 litres
D   11 litres

24 The plan for the new bathroom is drawn to the scale of 1:50. The window is 1.3 metres wide. What is the measurement of the window on the drawing?

A 6.5mm
B 26.0mm
C 6.5cm
D 26.0cm

25 One wall of the bathroom is to be covered with tiles. The dimensions of the wall to be tiled are shown in the diagram.

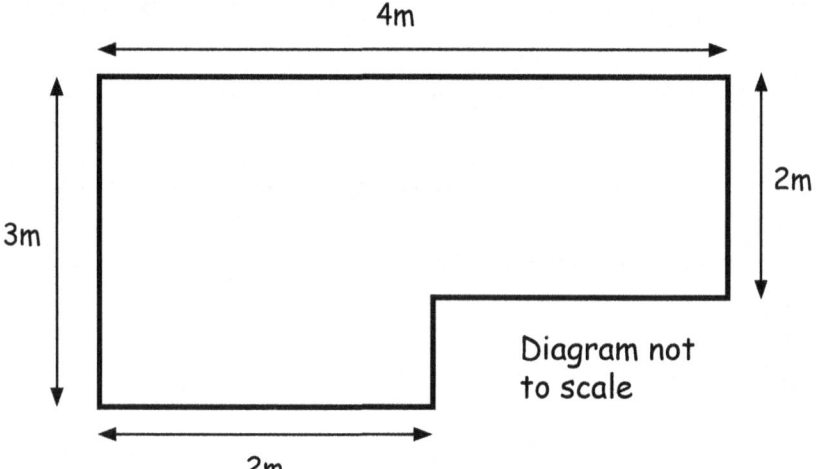

Diagram not to scale

Each tile costs £1 and measures 250mm by 250mm. What is the cost of the tiles required to cover the wall area?

A £40
B £80
C £96
D £160

26 A tiler is contracted to tile the bathroom wall. He charges a fixed call-out fee of £49.95 with an hourly rate of £8.95 and estimates that the job will take about 3 hours and 45 minutes. What is the best estimate of the cost of the job?

A (9 x 3) + 50
B (9 + 50) x 3
C (9 x 4) + 50
D (9 + 50) x 4

Questions 27 to 30 are about a mother and her daughter planning a holiday in Italy.

The table records highest and lowest temperatures in seven Italian cities.

Location	Lowest temperature °C	Highest temperature °C
Ancona	-3	36
Bari	-1	45
Florence	-11	39
Milan	-13	38
Naples	-2	49
Rome	-1	42
Venice	-8	38

27   What is the range of the Highest recorded temperatures?

    A   09°C
    B   12°C
    C   13°C
    D   14°C

28   What is the median of the Lowest recorded temperatures?

    A   -13°C
    B   -11°C
    C   -8°C
    D   -3°C

29   A mother and her daughter plan for a holiday in September.

September 2008 Calendar						
Sunday	Monday	Tuesday	Wednesday	Thursday	Friday	Saturday
1	2	3	4	5	6	7
8	9	10	11	12	13	14
15	16	17	18	19	20	21
22	23	24	25	26	27	28
29	30					

The brochure prices for 3 four-star hotels in Venice are shown:

2008 Holiday Prices (£ per person)						
Hotel	Firenze		Livorno		Rialto	
Number of Days	7	10	7	10	7	10
Departure day	Mon	Wed	Thurs	Fri	Sat	Sun
Pricing period:						
31 May – 14 Jun	495	595	529	775	570	810
15 Jun – 28 Jun	505	605	539	785	590	820
29 Jun – 10 Jul	515	615	549	795	610	830
11 Jul – 31 Jul	525	645	559	805	640	840
1 Aug – 14 Aug	545	665	580	820	660	860
15 Aug – 27 Aug	580	680	600	840	680	880
28 Aug – 10 Sep	560	665	550	870	690	840
11 Sep – 30 Sep	525	625	539	785	680	820

What is the latest departure date for the holidaymakers for a 10-day holiday to the Firenze hotel during September?

A   22nd September
B   23rd September
C   24th September
D   25th September

30   How much would the mother and her daughter save in total by choosing a 10-day holiday to the Firenze hotel during the pricing period 31st May to 14th June rather than at the most expensive pricing period for the same hotel?

A   £85
B   £140
C   £170
D   £370

Questions 31 to 33 are about the monitoring a patient's temperature levels after admission to hospital.

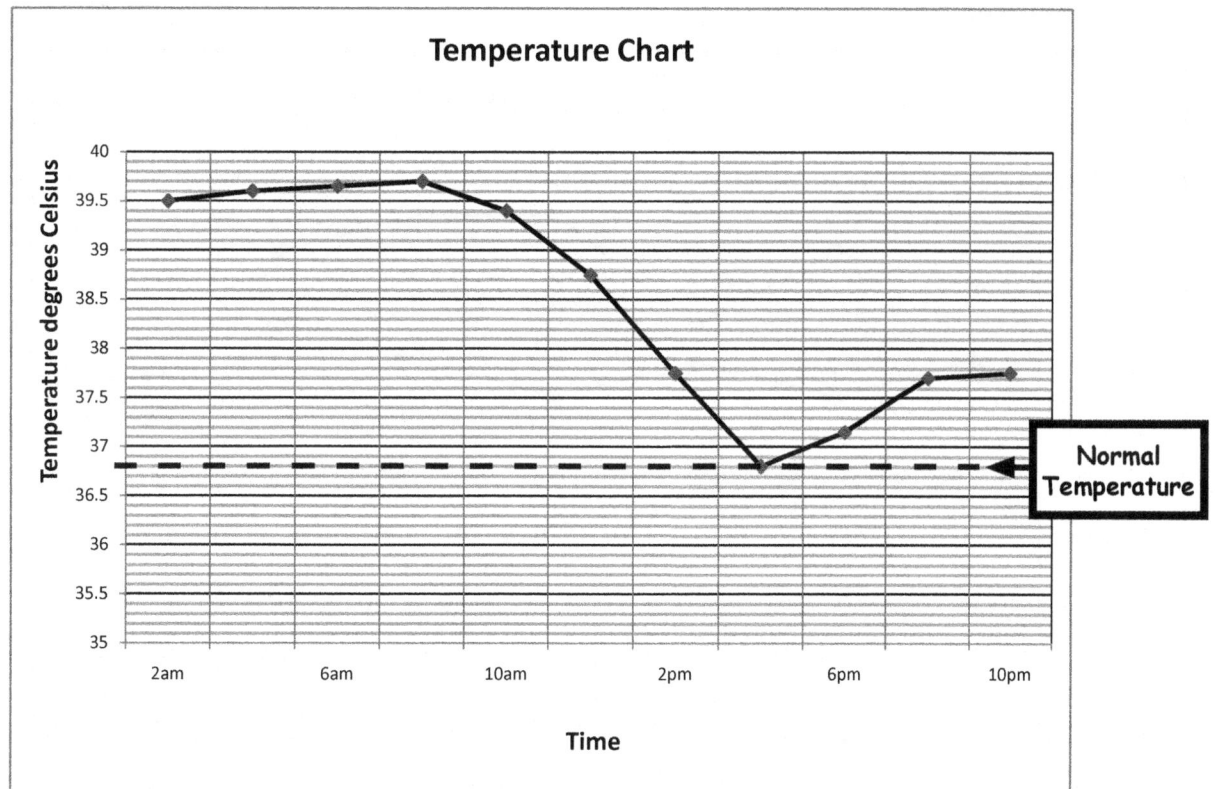

31   What was the patient's temperature around midday, rounded to the nearest 0.1°C?

   A   37.7°C
   B   38.8°C
   C   39.1°C
   D   39.6°C

32   The patient was admitted at 1am. How long did it take for her temperature to reach normal level?

   A   8 hours
   B   9 hours
   C   11 hours
   D   15 hours

33   Over the time period shown on the chart, which figure is the closest estimate of the range of temperatures recorded for this patient?

   A   2.3°C
   B   2.5°C
   C   2.7°C
   D   2.9°C

Questions 34 and 35 are about chocolate.

A chocolate manufacturing factory supplies chocolate to confectionery shops.

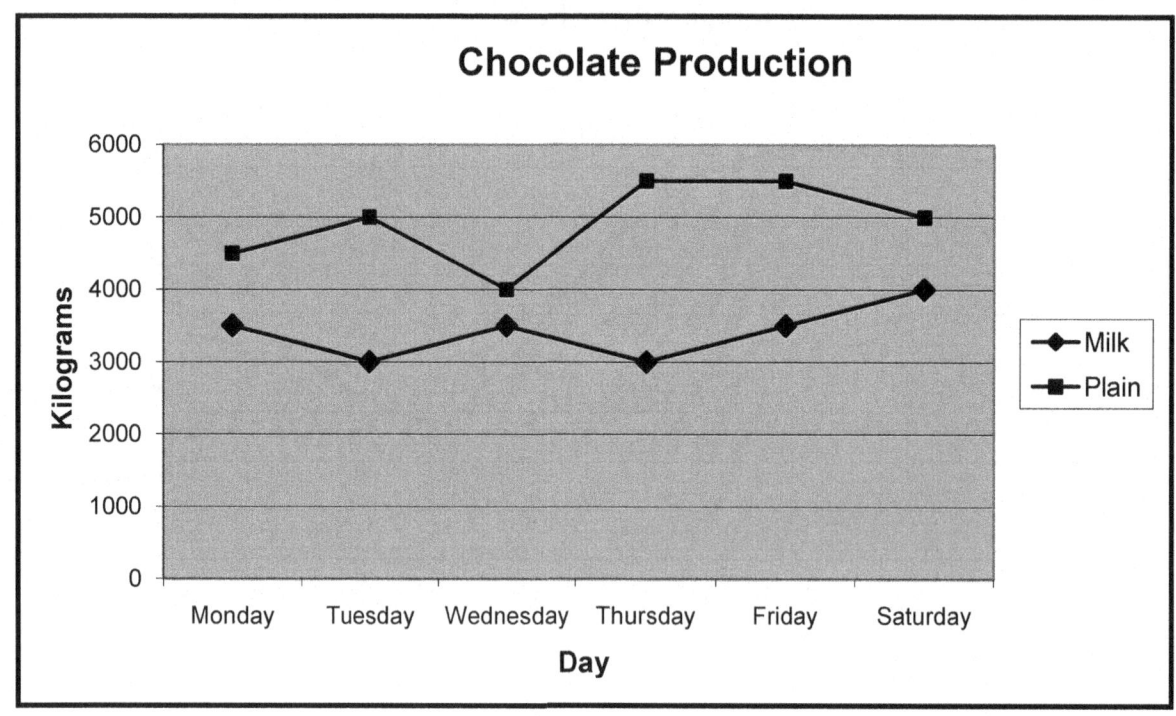

34 The top selling varieties are finest quality plain chocolate and milk chocolate. How many kilograms were produced of both milk and plain chocolate on Friday?

A   2500kg
B   5500kg
C   6500kg
D   9000kg

35 To promote the chocolate brand, a model coach is to be constructed from chocolate. Two solid slabs of chocolate form the chassis of the chocolate model coach. The dimensions are shown in the diagram.

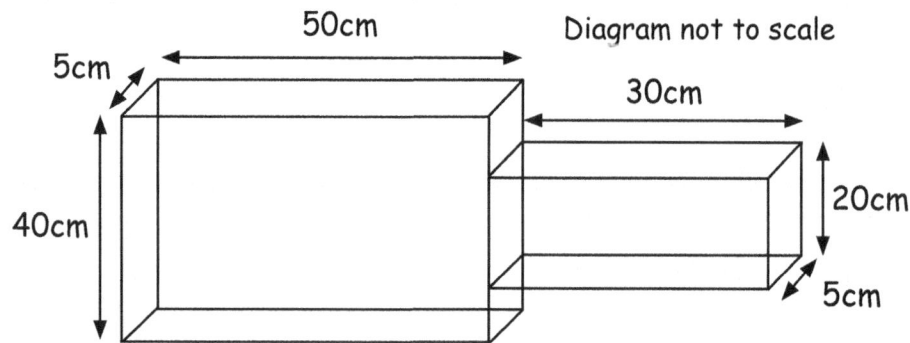

What volume of chocolate, in cubic metres, is required to make the chocolate model chassis?

A   13m³
B   1.3m³
C   0.13m³
D   0.013m³

Questions 36 and 37 are about two spa hotels in a national hospitality chain.

A marketing officer plots charts and a graph to show a comparison of profit levels for two hotels over a five-year period using the data in the table:

Hotel Profit / Loss York and Durham					
Year	1	2	3	4	5
York Manor (£000)	2030	3110	4250	5670	6890
Durham Oak (£000)	-600	-400	1001	6080	10300

Chart 1

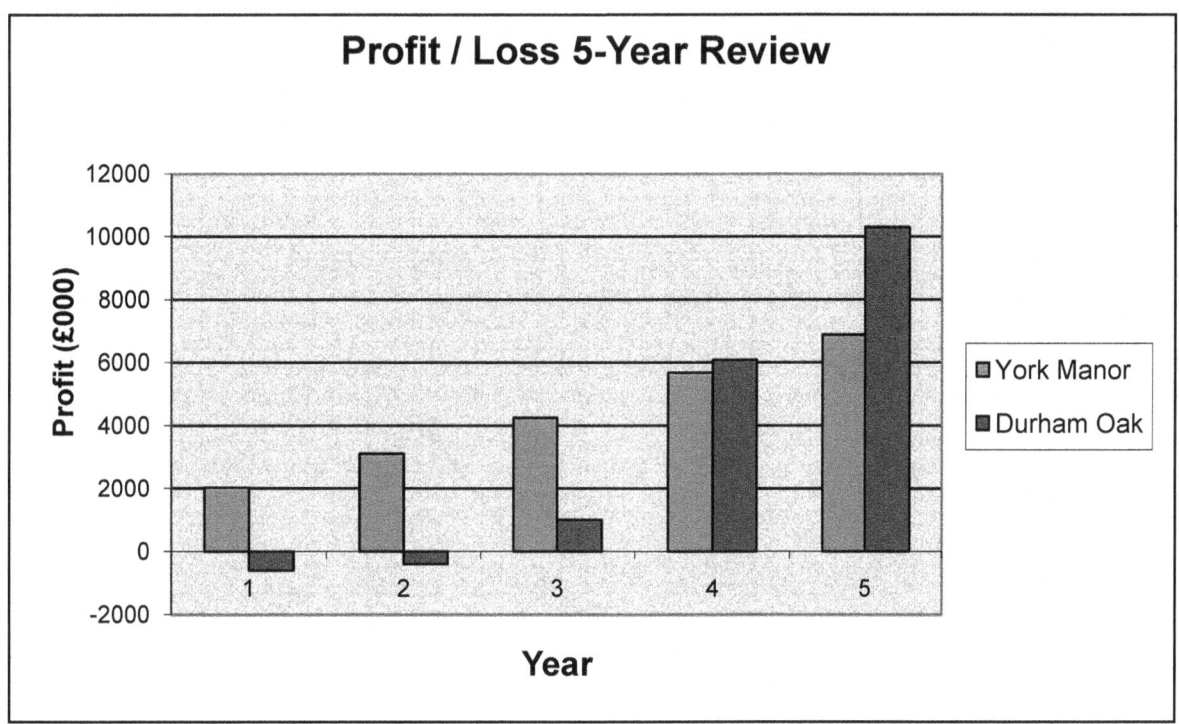

Chart 2

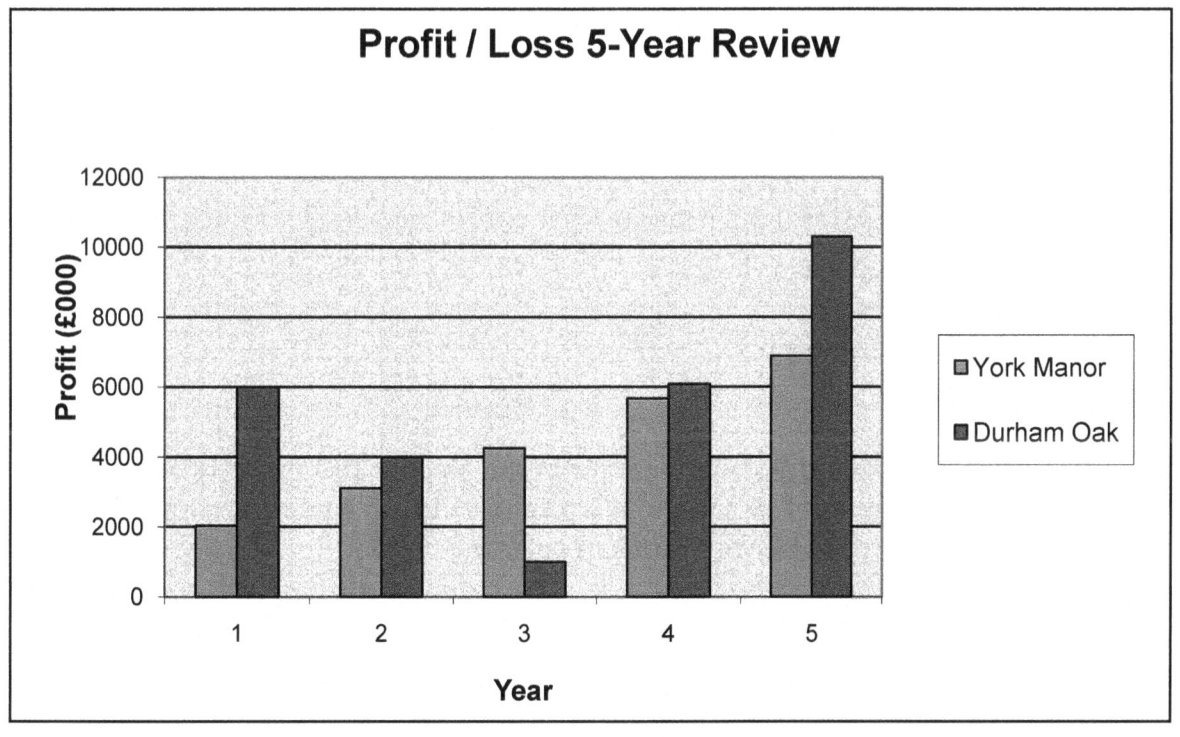

Chart 3

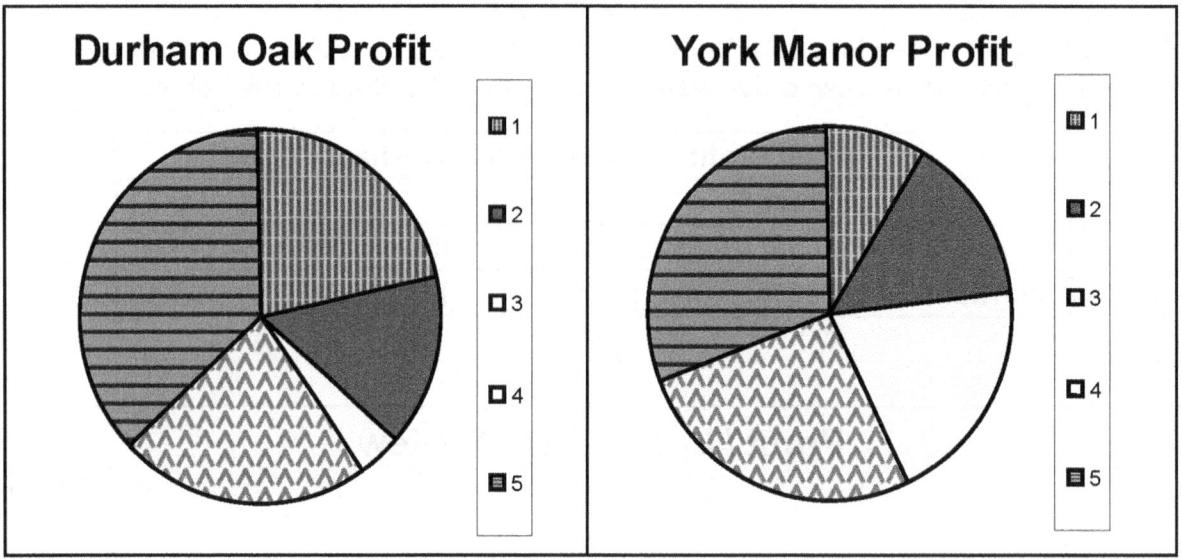

Graph 1

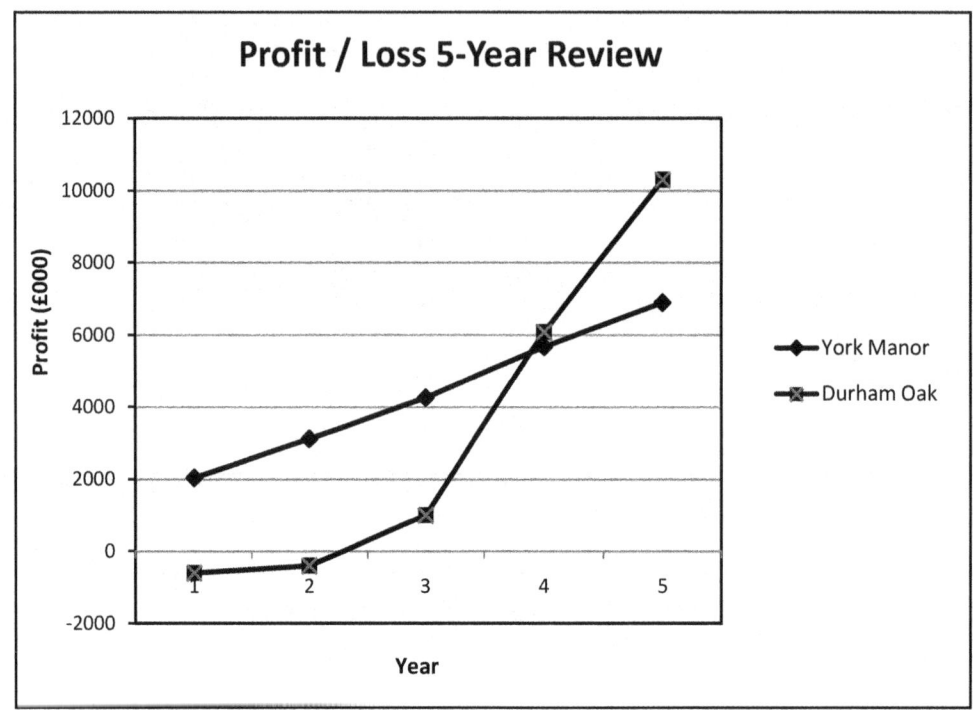

36   Which of the illustrations represent the information most accurately?

   A   Chart 1 and Graph 1
   B   Charts 2 and 3
   C   Chart 3 and Graph 1
   D   Charts 1 and 2

37   The Durham Oak and York Manor together employ 153 staff in total.
     Out of the total staff, 92 work in catering roles.   What fraction,
     approximately, of employees work in catering roles?

   A   1/5
   B   2/5
   C   3/5
   D   4/5

Questions 38 to 40 are about airline baggage allowances.

The bar chart show the number of times in one day when passengers were surcharged for exceeding the weight of baggage allowed by a European airline.

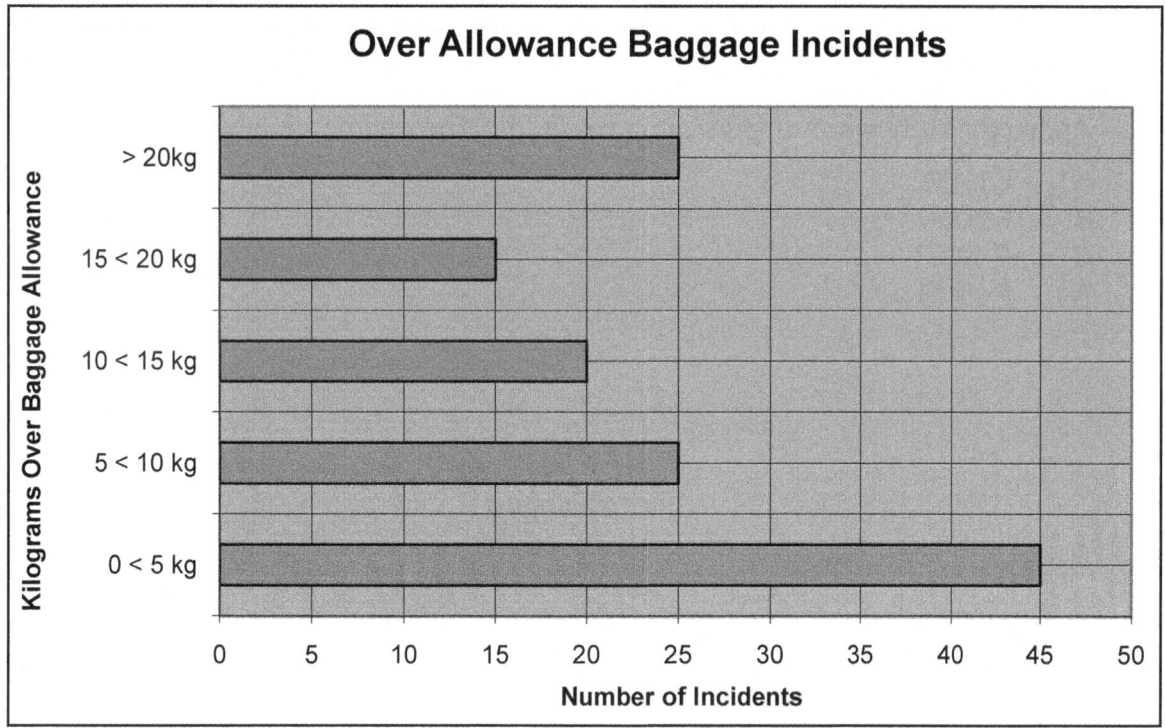

38 How many incidents in total were there of baggage exceeding the weight allowance by **less than** 20 kilograms?

A   15
B   55
C   60
D   105

39 The table shows the charges, in Euros, for excess baggage, per kilogram over the permitted weight allowance.

Excess Baggage Charges per kilogram (€)	
Less than 20kg	5
More than 20kg	10

In one week:
- 396 passengers were charged €5 per kilogram of overweight baggage;
- 99 passengers were charged €10 per kilogram of overweight baggage.

Approximately what percentage of passengers in one week paid the higher rate for excess baggage?

A   25%
B   20%
C   15%
D   10%

40 In one day, the airline received €3201 in excess baggage charges.

> ## The exchange rate is 1.6 Euros to the British Pound

Approximately what amount, in pounds, did the airline receive?

A   £1600
B   £2000
C   £4900
D   £6500

End of Paper

Practice Multiple-choice Paper
Suitable for:

Key Skills Level 2 Application of Number
Level 2 Adult Numeracy

Paper Four

YOU NEED

- This test paper.
- A pen.
- A pencil and eraser.
- An Answer Sheet.
- A ruler marked in centimetres and millimetres

You may NOT use a calculator.
You may use a bilingual dictionary.
There are 40 questions on this paper. Try to answer ALL the questions.
When you have completed the questions you must check your answers, then check them again.

YOU HAVE QUARTER OF AN HOUR TO READ THE PAPER
AND ONE HOUR TO COMPLETE THE 40 QUESTIONS

INSTRUCTIONS

- Make sure you write your name and today's date on the Answer Sheet. Use a pen to do this.

- Use a pencil to mark your answers so if you change your mind you can erase your choice and select another.

- Make sure that for each question you have only selected one answer. If you select more than one, the answer will not be marked.

- Read each question carefully before you select an answer.

Note for learners and tutors: This is a practice test that has been designed to closely resemble the questions and question styles of a "live" paper.

Questions 1 to 5 are about a holiday to Spain.

The chart shows typical monthly rainfall recorded over six months for two Spanish resorts, Sitges and Santander.

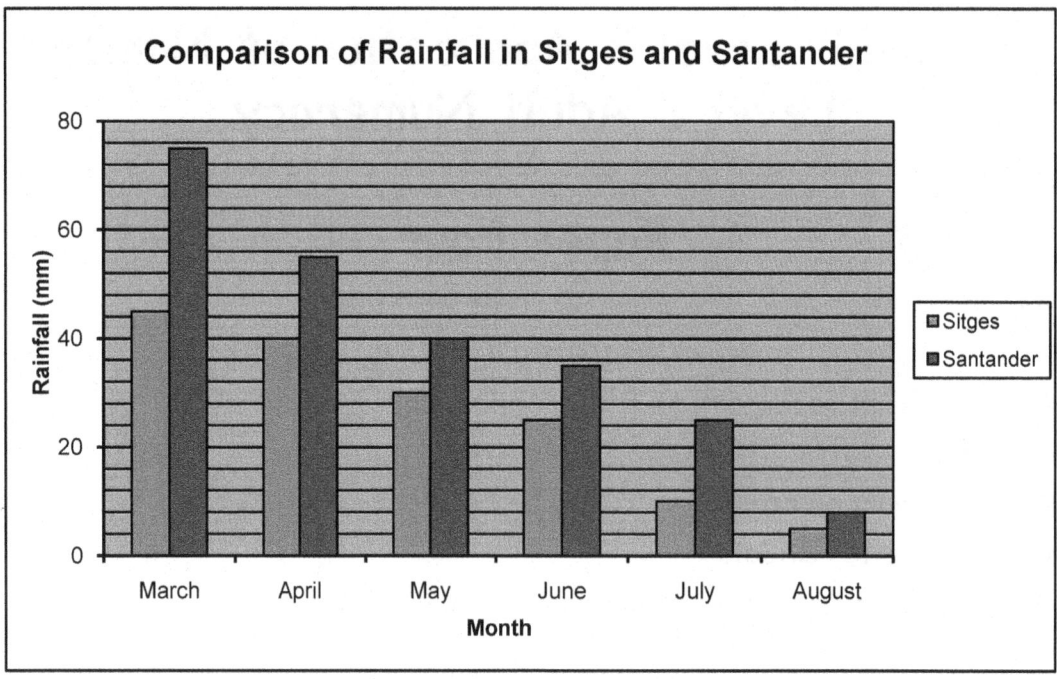

1   Which of the following statements is true?

   A   Sitges has more rainfall than Santander in total.
   B   Sitges has most rainfall in July but is drier than Santander in August.
   C   Santander is drier than Sitges every month.
   D   The range of monthly rainfall is greatest for Santander than for Sitges.

2   A young couple travelling to Sitges plan to take £400 spending money. Their bank does not charge commission on foreign currency exchanges.

     The current exchange rate gives 1.6 Euros to the pound.

     How many Euros will the couple receive for their £400?

   A   €640
   B   €624
   C   €610
   D   €600

3   The couple travel by car from the UK. On their way through Spain, they see a road sign showing the distance to Sitges:

     1 mile = 1.6 kilometres

     How far is 64 kilometres in miles?

   A   102.4 miles
   B   40 miles
   C   6 miles
   D   4 miles

Sitges 64 km

4   The fuel tank holds 32 litres when full.

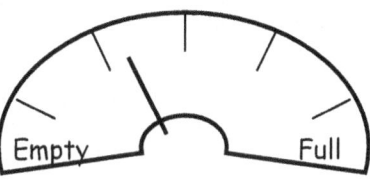

What is the best estimate of how much petrol is left in the tank to the nearest litre?

A   16 litres
B   12 litres
C   10 litres
D   8 litres

5   The couple stop for petrol when they reach Sitges. They buy 30 litres of petrol. One litre of petrol costs €1.47. How many Euros do they spend on petrol?

A   €30.00
B   €31.25
C   €37.50
D   €44.10

Questions 6 to 14 are about a refreshments kiosk in a local cinema complex.

The chart shows the sales income from soft drinks in one week.

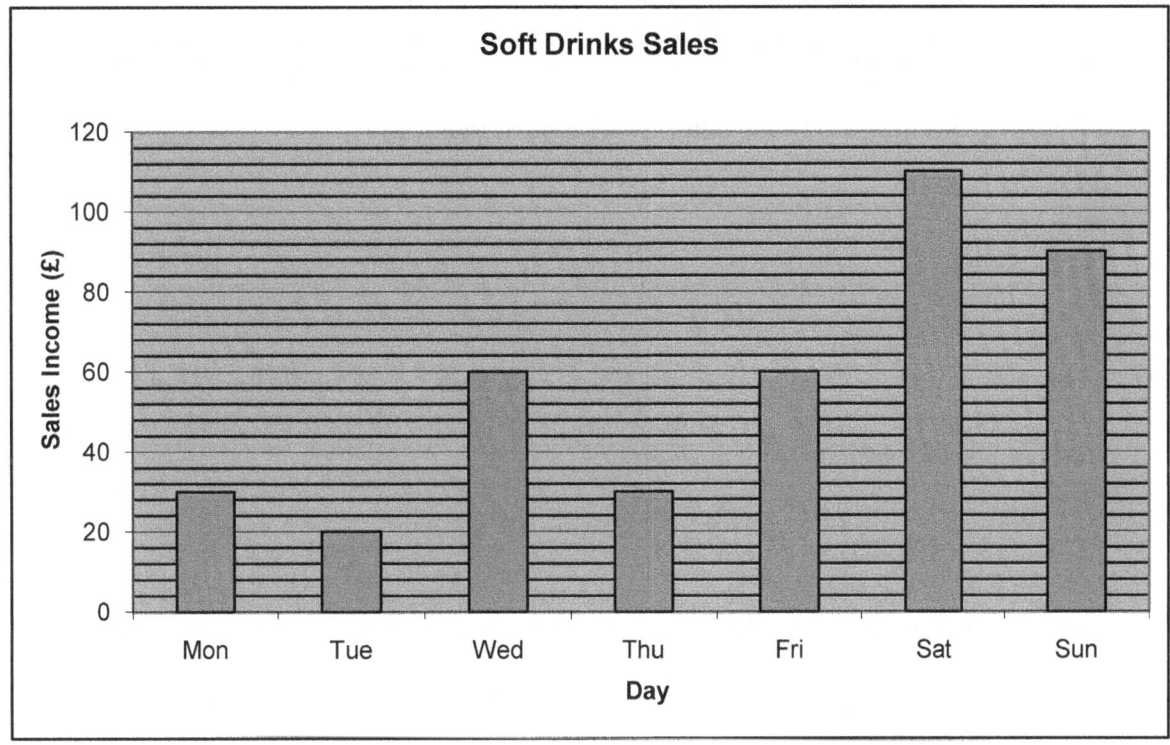

6   How much is the total income from soft drinks during the whole week?

A   £110
B   £200
C   £400
D   £500

7   Looking at the Soft Drinks Sales chart, which statement is accurate?

The income from soft drinks is ...?

A   highest during the middle of the week
B   highest on Sunday
C   the same on Monday and Tuesday
D   the same for Saturday and Sunday as all of the weekdays added
     together

8   What is the mean income from soft drinks per day, to the nearest pound?

A   £35
B   £42
C   £56
D   £57

9   The table shows the number of hours worked by 4 cinema attendants in a
week:

Attendant	Monday	Tuesday	Wed	Thursday	Friday	Saturday	Sunday
Polly	7 hrs	0 hrs	0 hrs	7 hrs	8 hrs	7 hrs	6 hrs
Sally	8 hrs	7 hrs	7 hrs	9 hrs	7 hrs	0 hrs	0 hrs
Billy	0 hrs	7 hrs	0 hrs	6 hrs	7 hrs	7 hrs	9 hrs
Phil	6 hrs	7 hrs	8 hrs	0 hrs	0 hrs	10 hrs	9 hrs

What is the mode of the number of hours worked?

A   6 hrs
B   7 hrs
C   8 hrs
D   10 hrs

10   The cinema attendants are paid £6.30 per hour for weekday hours worked.
The hourly-rate during weekend shifts is £9.15. Which calculation shows
how much Phil earns for the week?

A   (21 x £6.30) + (19 x £9.15)
B   (19 x £6.30) + (21 x £9.15)
C   (£9.15 + £6.30) x 21
D   (£9.15 + £6.30) x 19

Please go on to the next page

11   The kiosk assistant weighs out a standard bucket of toffee popcorn.

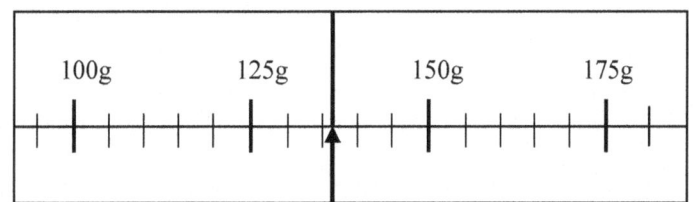

What is the weight of popcorn that the customer receives, to the nearest 1g?

A   122g
B   127g
C   132g
D   137g

The diagram shows one of the tubs of ice cream for sale at the kiosk.

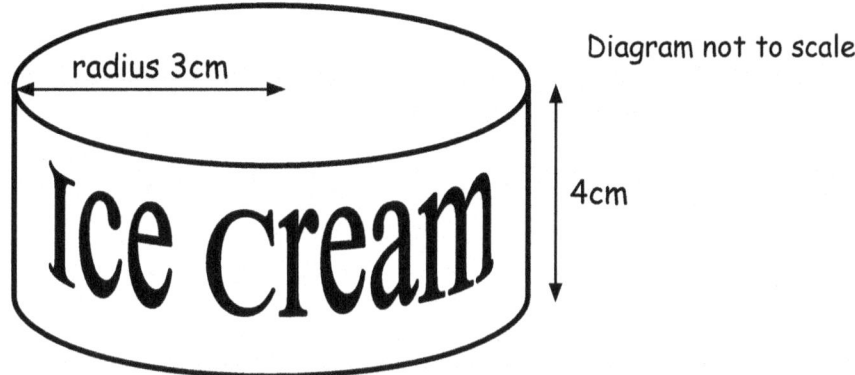

The tub is filled to the rim and sealed with a waxed paper lid.

The formula to work out the volume of cylinder is pi times radius squared times height.

$$\pi \times r^2 \times h \quad \text{(the approximate value of pi } (\pi) = 3)$$

12   What volume of ice cream, in square centimetres, is contained in the tub?

A   84cm³
B   108cm³
C   144cm³
D   216cm³

Please go on to the next page

13 How many tubs of Ice Cream will fit into the container shown in the diagram?

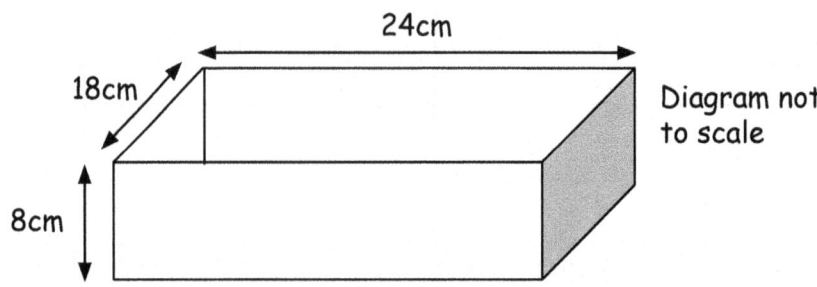

24cm

18cm

Diagram not to scale

8cm

A 12
B 18
C 24
D 30

The temperature of the ice cream at the point-of-sale is -6° Celsius. The temperature in the cinema is 22° Celsius (°C).

14 What is the difference between the temperature in the cinema and the ice cream on sale, in degrees Celsius?

A 26°C
B 28°C
C 29°C
D 30°C

Questions 15 to 20 are about a regional inter-schools sports event.

The table shows goals scored by 20 teams in the first round of a football competition.

Goals scored				
0	5	1	0	3
1	1	3	2	4
2	0	1	1	3
2	2	4	2	1

15 What percentage of teams scored **less than** 2 goals?

A 9%
B 20%
C 40%
D 45%

16 What is the **range** of scores in the first round of the competition?

A 2
B 4
C 5
D 7

17 What is the **median** number of goals scored in the competition?

    A   2
    B   3
    C   4
    D   5

In the final round of the lower schools' high jump competition, the height jumped by each high jumper is measured to decide the medal places.

The diagram shows part of the measure rule and the arrow marks the height jumped by the silver medalist.

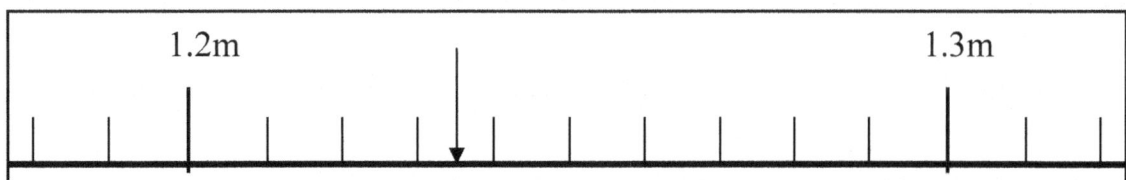

18 The **winner** of the gold medal jumped a **further** 6.0cm. What is the nearest estimate of the height of the **winning** jump?

    A   1.165m
    B   1.235m
    C   1.275m
    D   1.295m

19 Nine high jumpers took part in the competition final and each jumped only once. The mean height of their jumps is 1.225m. Which statement is true?

    A   most people jumped further than 1.225m
    B   the total height jumped is 11.025m
    C   half of the competitors jumped further than 1.225m
    D   everyone jumped further than 1.225m

20 A mini-bus is hired to transport competitors from one school to the inter-schools competition venue. The mini-bus hire company charge a fixed fee of £30 per day plus £15 for every hour hired.

The mini-bus is hired out from 8.30am to 4.30pm.

Which calculation shows how to work out the total mini-bus hire costs?

    A   (£30 + £15) x 7.5
    B   (£30 + £15) x 8.0
    C   (£30 x 8.0) + £15
    D   (£15 x 8.0) + £30

Please go on to the next page

Questions 21 to 23 are about choosing a mobile phone network account.

The table shows details of three mobile network deals.

Network	Handset	Monthly Service Rental	Hands free Kit	Optional Monthly Insurance
Apple Net - Crystal	£49.99	£19.99	£10	2.49
Fidophone - Blue	£39.99	£29.99	£5	3.49
Moto-Mobile - Flip	£29.99	£29.99	Free	3.49

Alison chooses the Apple Net - Crystal package with the hands-free kit and pays a year's line rental in advance, without insurance.

21   What is the total cost of Alison's choice of mobile phone network?

A   £80.00
B   £82.49
C   £299.87
D   £389.87

22   Philip chooses the Moto-Mobile package, pays for one year's Service Rental and insurance. The total is £431.75. Which calculation can he use to check his bill?

A   £431.75 - (12 × £29.99) - (12 × 3.49) = £29.99
B   £431.75 + (12 × £29.99) - (12 × 3.49) = £29.99
C   £431.75 - (12 × £29.99) + (12 × 3.49) = £29.99
D   £431.75 + (12 × £29.99) + (12 × 3.49) = £29.99

23   Moto-Mobile calls are charged by the second. A 2-minute call costs 20p. To the nearest penny, what is the cost of a call lasting 270 seconds?

A   135p
B   90p
C   60p
D   45p

Questions 24 to 26 are about Council Tax charges.

The table shows the Council Tax rate set for properties in each valuation band.

Valuation Band	A	B	C	D	E	F	G	H
Council Tax £s	765	877	989	1122	1347	1592	1847	2235

A discount of 25% is allowed for single occupancy of a property.

24   How much Council Tax does a single occupant have to pay in a property rated in band F?

A   £398.00
B   £592.25
C   £796.00
D   £1194.00

25 Council Tax rates will increase next year by 6.7%. Which calculation can be used to work out the increased rate for a house in band E?

A £1347 × $\frac{6.7}{100}$

B £1347 + $\frac{6.7}{100}$

C £1347 + $\left(£1347 × \frac{6.7}{100} \right)$

D £1347 × $\left(£1347 × \frac{6.7}{100} \right)$

26 What is the range of Council Tax charges?

A £1234.50
B £1470.00
C £1847.00
D £2235.00

Questions 27 to 32 are about a hotel spa treatment room.

The scale drawing shows the plan view of a hydrotherapy treatment room.

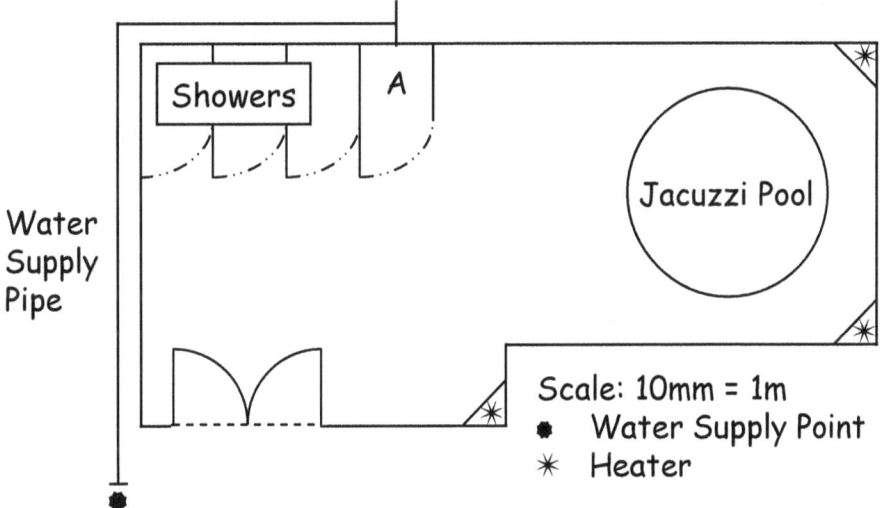

27 What is the total floor area of the treatment room?

A 20.0m²
B 25.0m²
C 45.4m²
D 50.5m²

28 The water supply pipe is installed around the outside of the treatment room as far as the fixing point in shower cubicle **A**. What is the length of the water supply pipe?

A 5.5m
B 6.9m
C 9.9m
D 10.5m

The Jacuzzi pool in the treatment room is 2.8m in diameter.

> The area of a circle is approximately $3 \times r^2$ (r = radius)

29 What is the area of the floor under the Jacuzzi pool?

    A $1.4m^2$
    B $1.96m^2$
    C $3.0m^2$
    D $5.88m^2$

30 The spa treatments are available to hotel guests seven days a week. The electric heaters in the treatment room are switched on at 8am and switched off at 8pm each day.

Each unit of electricity costs ten pence. Each of the three heaters uses 2 units of electricity every hour.

How much does it cost to heat the treatment room in a week, in pounds and pence?

    A   £50.00
    B   £50.40
    C   £54.00
    D   £72.00

31 Spa attendants work 6-hour shifts. Each attendant spends one hour cleaning the treatment room and shower cubicles. What percentage of an attendant's shift is spent cleaning, to the nearest whole number?

    A   1.6%
    B   1.67%
    C   16.7%
    D   17%

32 The table shows the number of hours booked in the treatment room by 21 guests at the hotel during one week.

Treatment hours booked	7, 4, 5, 3, 4, 3, 6, 8, 7, 9, 5, 4, 6, 3, 8, 3, 5, 3, 5, 3, 4

What is the mean number of hours spent by guests in the treatment room in a week?

    A   3 hours
    B   4 hours
    C   5 hours
    D   6 hours

The table shows the air temperatures for one day in February in four cities in different parts of England.

Temperature recorded at time:	Birmingham	Kent	Yeovil	York
Midnight °C	2.6	-1.5	2.6	-1.9
Dawn °C	-0.9	-4.3	-0.8	-4.0

33 Which city experienced the greatest drop in over-night temperature?

    A   Birmingham
    B   Kent
    C   Yeovil
    D   York

Questions 34 and 35 are about the height of trees.

The lines on the diagram show the relative height of two trees, A and B.

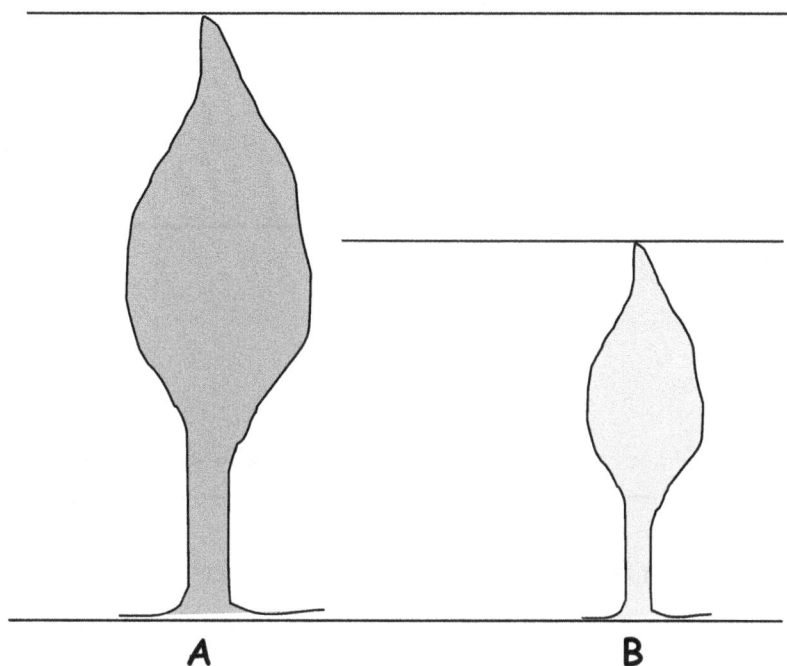

34 Approximately what fraction of the height of tree A is the smaller tree, B?

    A $\frac{1}{2}$

    B $\frac{5}{8}$

    C $\frac{2}{3}$

    D $\frac{3}{4}$

35 Tree A is 8 feet 6 inches high.

| 1 inch = 2 .54cm |

Which calculation can be used to work out the height of the tree in centimetres?

A (8 x 12 + 6) ÷ 2.54
B (8 x 12 - 6) x 2.54
C (8 x 12 x 6) ÷ 2.54
D (8 x 12 + 6) x 2.54

Questions 36 to 38 are about a sea trip, organised by a bird-watching club, to observe and photograph wild sea birds.

36 The line on the plan shows the distances from the coast to the Farne Isle.

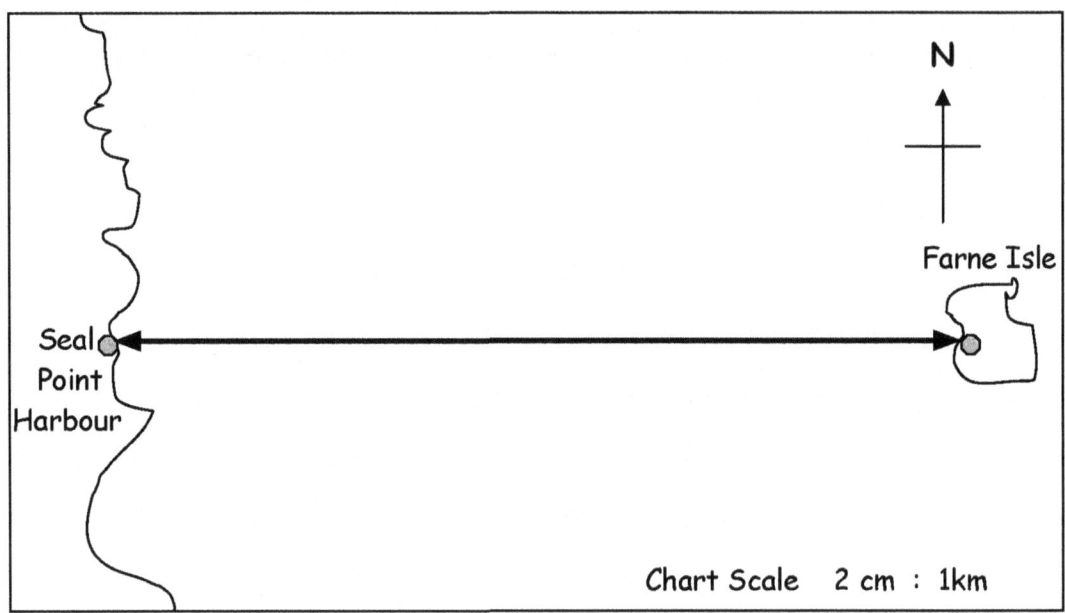

How far is it from Seal Point Harbour to the Farne Isle, to the nearest kilometre?

A 11.6km
B 11.2km
C 5.8km
D 5.2km

Please go on to the next page

37 The bird-watching club hires a boat to cross to the isle. Club members have from 10.00am to 3.00pm.

> Boat Hire: £40 plus £12 per hour

How much does it cost the club to hire the boat?

A £52
B £100
C £124
D £144

38 Members of the bird-watching club qualify for a 20% discount on all photographic printing with a local printing firm.

Photographic Service (per image)	Cost
Digital Processing	25p
Standard Size Printing	21p
Enlarged Size Printing	28p
Display Mount	70p

A club member has 23 images of sea birds photographed on the Farne Isle; 20 of these require standard printing and 3 images need to be enlarged and display mounted for entry in a photographic competition.

What is the total cost to process his 23 images?

A £12.89
B £12.69
C £10.31
D £10.14

39 The Standard Printing size is 13cm x 9cm. The Standard Printing size in comparison to the Enlarged Printing size is in the ratio 2 : 3. What is the length and width of the Enlarged Printing size?

A 10cm x 6cm
B 15cm x 12cm
C 19.5cm x 13.5cms
D 26cm x 27cms

Please go on to the next page

40 The diagram shows four areas of land in a market garden.

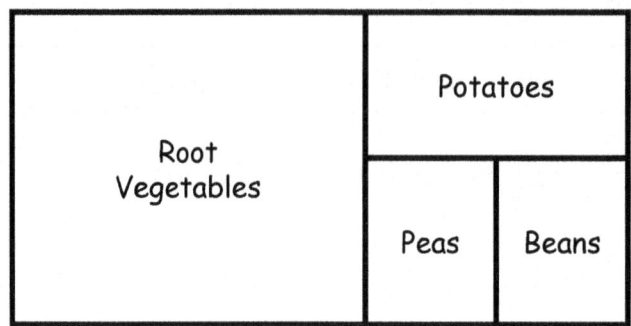

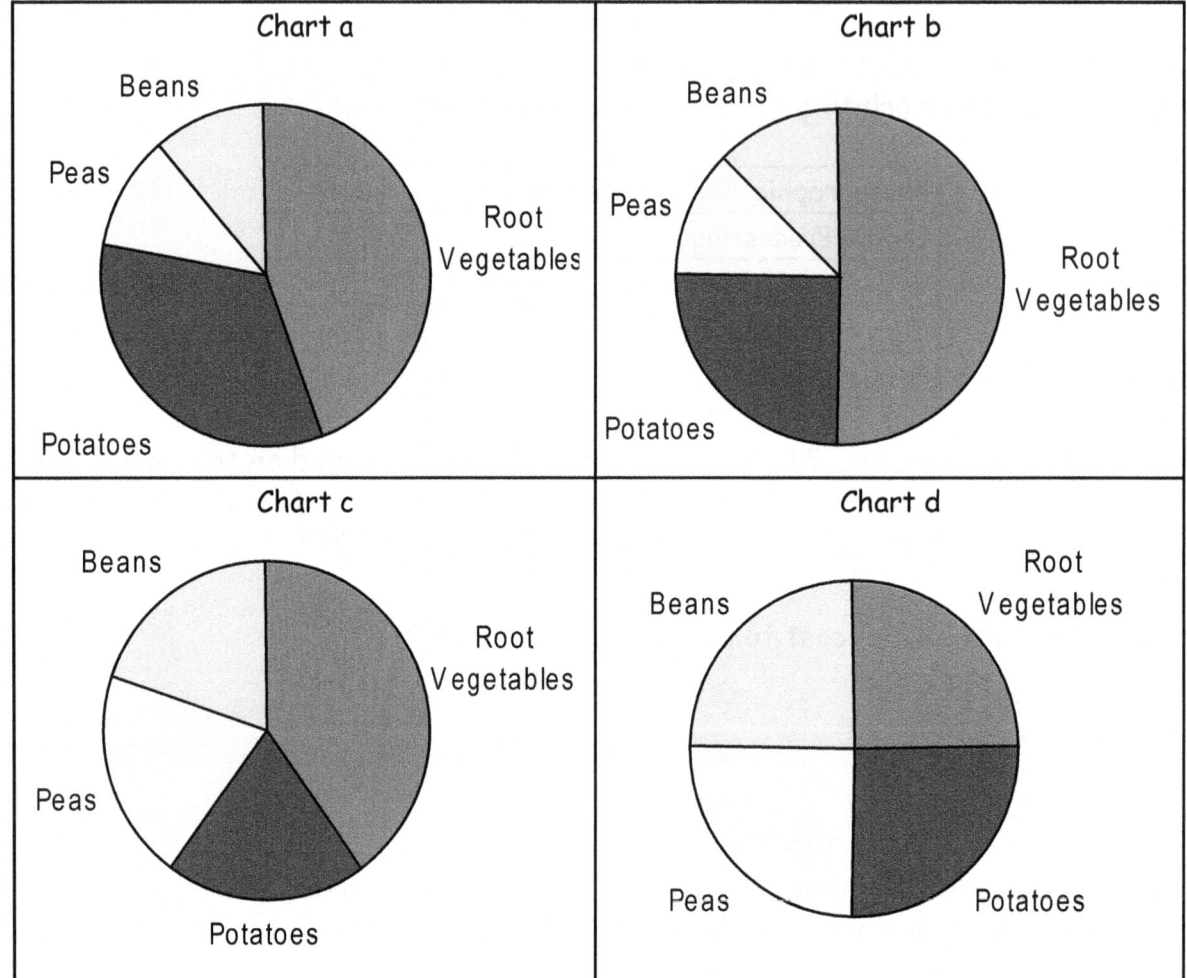

Which chart most accurately represents the four areas of land?

A   Chart a
B   Chart b
C   Chart c
D   Chart d

End of Paper

Practice Multiple-choice Paper
Suitable for:

Key Skills Level 2 Application of Number
Level 2 Adult Numeracy

Paper Five

YOU NEED

- This test paper.
- A pen.
- A pencil and eraser.
- An Answer Sheet.
- A ruler marked in centimetres and millimetres

You may NOT use a calculator.
You may use a bilingual dictionary.
There are 40 questions on this paper. Try to answer ALL the questions.
When you have completed the questions you must check your answers, then check them again.

YOU HAVE QUARTER OF AN HOUR TO READ THE PAPER
AND ONE HOUR TO COMPLETE THE 40 QUESTIONS

INSTRUCTIONS

- Make sure you write your name and today's date on the Answer Sheet. Use a pen to do this.
- Use a pencil to mark your answers so if you change your mind you can erase your choice and select another.
- Make sure that for each question you have only selected one answer. If you select more than one, the answer will not be marked.
- Read each question carefully before you select an answer.

Note for learners and tutors: This is a practice test that has been designed to closely resemble the questions and question styles of a "live" paper.

Questions 1 to 4 are about employees in a call-centre.

1   A personnel officer gathers employee data on name, age, gender and wages. The call-centre currently has 426 women and 214 men working there. Approximately, what is the ratio of women to men?

A   3 : 1
B   2 : 1
C   1 : 3
D   1 : 2

2   For annual statistics the personnel officer wants to compare employee ages and weekly earnings. Which method is most effective to present the information?

A   a bar chart
B   a pie chart
C   a line graph
D   a scatter graph

3   The personnel officer analyses employee ages and finds that out of 650 employees, 260 of them are under 30 years old. What fraction of employees is under 30 years old?

A $\dfrac{1}{5}$

B $\dfrac{2}{5}$

C $\dfrac{3}{5}$

D $\dfrac{4}{5}$

4   The personnel officer needs to find the modal (most common) age group of the call-centre's employees.

Which of the following is the most appropriate method of finding the modal age group?

A   draw a scatter graph with employees' ages grouped in intervals of ten
B   draw a frequency table with employees' ages grouped in intervals of ten
C   draw a line graph plotting each employee's age against their name
D   draw a bar chart plotting each employee's age against their name

Please go on to the next page

Questions 5 to 10 are about the organising of activities at a County show.

The organising committee of the County show places an order with a printer for 2000 leaflets to publicise the event.

A printer uses a formula, (that includes a fixed fee plus the cost per leaflet), to calculate the total cost of printing:

> ## C = £17 + (5p × n)
>
> C is the total cost (in £s); n is = number of leaflets

5   What is the total cost of printing the leaflets?

- **A**   £17. 05
- **B**   £34.10
- **C**   £100.00
- **D**   £117.00

6   Name badges are required to identify members of the organising committee on the day of the show. Jane volunteers to make the circular badges, measuring 80 millimetres in diameter, from sheets of card.

From a sheet of card measuring 24cm by 40cm, how many name badges can be made?

- **A**   40
- **B**   24
- **C**   15
- **D**   12

Ann volunteers to make scones for the cream teas to be sold from a stall in the refreshment tent. The recipe for the scones uses the following ingredients to make 12 fruit scones.

> ## Scones Recipe
>
> 250g plain flour
> 60g butter
> 175ml skimmed milk
> 2 teaspoons of baking powder
> 50g sultanas

Twelve batches of scones are baked.

7   How many 250g packs of butter does Ann buy to make the twelve batches of scones?

- **A**   2
- **B**   3
- **C**   4
- **D**   5

8 Traditional cream teas are sold at £2.25 each. 135 cream teas are served. The cash box contains £303.75. Which calculation can be used to check this amount?

A $\dfrac{303.75}{135}$

B $\dfrac{135}{303.75}$

C $\dfrac{135}{225}$

D $\dfrac{135}{2.25}$

9 A prize stall gives a small prize to all ticket numbers ending in a three or a seven between 1 and 299. How many prize-winning tickets are there in total?

A 27
B 36
C 54
D 60

The amount of money raised by various stalls at the County show event is shown in the table:

Cream Teas	Cakes and Pastries	Candy and Chocolate	Ploughman's Lunches	Devil's Kitchen	Hot Dogs and Burgers	Prize Draw
£303.75	£128.20	£97.88	£171.50	£88.50	£145.75	£330.75

10 What is the range of these amounts?

A £242.25
B £215.25
C £186.38
D £88.50

Please go on to the next page

Questions 11 to 13 are about a young working mother's finances.

Linda works part-time and earns £8064 per year and is paid in 12 equal monthly salary payments.

The pie chart shows where Linda's money goes each month:

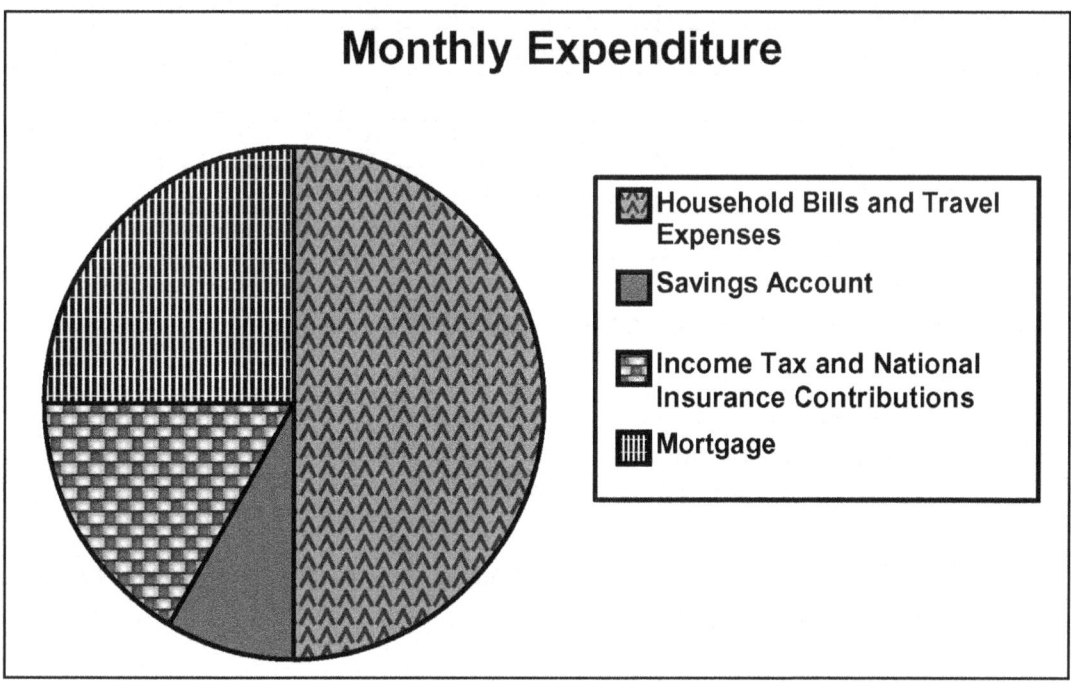

Monthly Expenditure

Household Bills and Travel Expenses

Savings Account

Income Tax and National Insurance Contributions

Mortgage

11   What fraction of Linda's salary does she put into the savings account?

    A $\frac{1}{30}$

    B $\frac{1}{12}$

    C $\frac{1}{8}$

    D $\frac{3}{10}$

12   How much of Linda's monthly salary is deducted for Tax and National Insurance Contributions?

    A   £22.40
    B   £56.00
    C   £84.00
    D   £112.00

13   Linda spends £47 each month on a bus pass to travel to and from work. If the price of the monthly bus pass stays the same, how much will Linda spend on travelling to work in three years?

    A   £1692
    B   £1410
    C   £1128
    D   £564

Questions 14 to 16 are about fund raising cycling events.

Ralph takes part in six 20km charity cycling races one summer season. His times are recorded in the table:

Date	Location	Minutes	Seconds
May 25	Kent	52	15
May 31	Harrow	49	25
June 04	Lincoln	47	49
June 11	Shrewsbury	50	26
July 24	York	56	36
August 14	Dorham	55	59

14   What is Ralph's mean time, in minutes and seconds?

   A   50 minutes 27 seconds
   B   52 minutes 15 seconds
   C   52 minutes 05 seconds
   D   55 minutes 59 seconds

15   The entry fee for each charity fundraising event is £9. A Spanish cyclist, Phillipe, enters the August event.

   > The exchange rate from British pounds to Euros is: £1 = €1.6

   How many Euros does Phillipe spend to enter the event?

   A   €9.60
   B   €10.60
   C   €14.40
   D   €15.00

16   A mini-bus and trailer, hired to transport entrants to the Dorham event, carries 22 passengers and costs £253.30 per day. The event organiser rounds up the numbers to the nearest ten to estimate the approximate cost per passenger.

   Which calculation gives the best estimate of the cost per passenger?

   A   260 ÷ 20 = £13.00
   B   250 ÷ 20 = £12.50
   C   250 ÷ 22 = £11.36
   D   260 ÷ 30 = £8.6

Please go on to the next page

The scale map shows a section of the cycle route at the Dorham event.

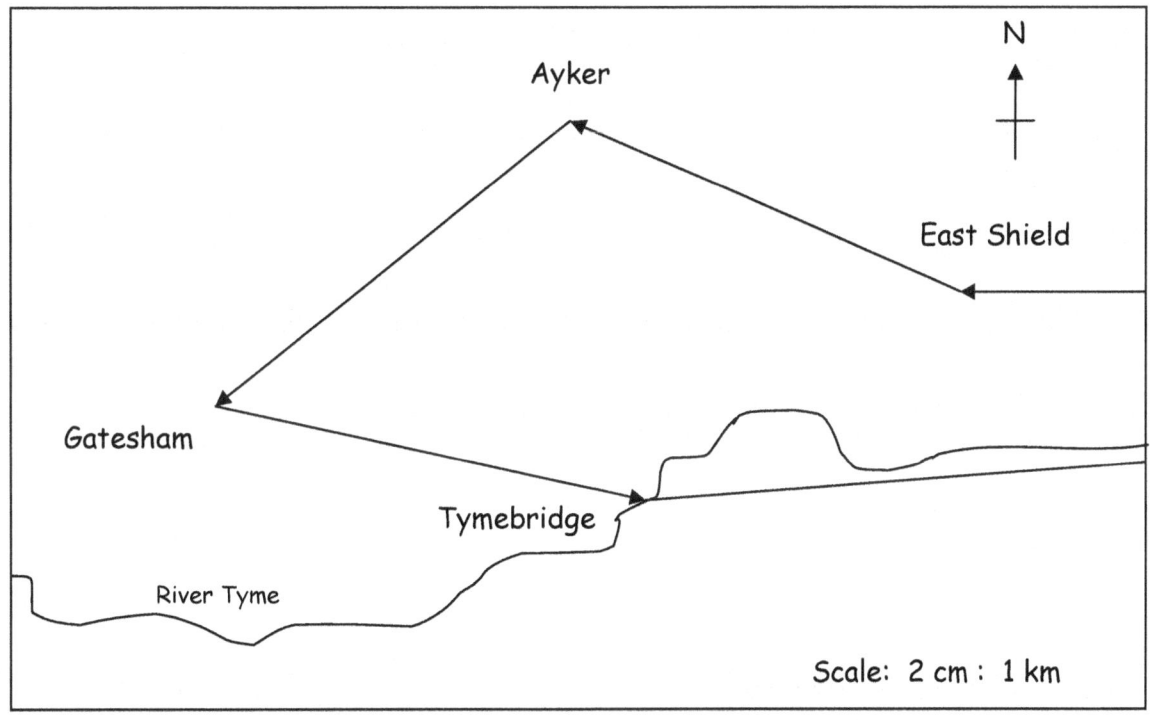

17   What is the distance, in kilometres, between East Shield and Gatesham?

A   23.80km
B   11.90km
C   5.95km
D   5.59km

Questions 18 to 21 are about the annual general meeting of Beauford's Football and Lawn Tennis club, held at Maston Hall Business Centre.

The room hire charges at Maston Hall are shown in the table.

Maston Hall Business Centre Private Hire Tariff		
	Westham Wing	Eastham Wing
Weekdays: per hour	£16	£12
Weekends: per hour	£24	£20
Conference Equipment Hire: per booking	£30	£24
Deposit required	20%	15%

18   Club Treasurer, Phillipa, approves the funds to pay the deposit for hiring the Eastham Wing for four hours on a Saturday including conference equipment. Which calculation can she use to work out the required deposit?

A $(20 + 24) \times 4 \times \dfrac{100}{15}$

B $(24 \times 4) + 20 \times \dfrac{100}{15}$

C $(20 + 24) \times 4 \times \dfrac{15}{100}$

D $(20 \times 4) + 24 \times \dfrac{15}{100}$

19 Club member, Jane, prepares 21 litres of orange squash to accompany the buffet lunch. Jane dilutes the orange squash concentrate with water in the ratio 2 : 5. How many litres of the orange squash concentrate does Jane use?

    A   7

    B   6

    C   5

    D   4

20 The club committee wants to raise the annual membership fee from £24 to £30.

What is the percentage increase in the annual membership fee?

    A   12%

    B   20%

    C   25%

    D   30%

A vote is taken to decide on which day of the week members would prefer to hold the monthly meeting.

The results are shown in the bar chart.

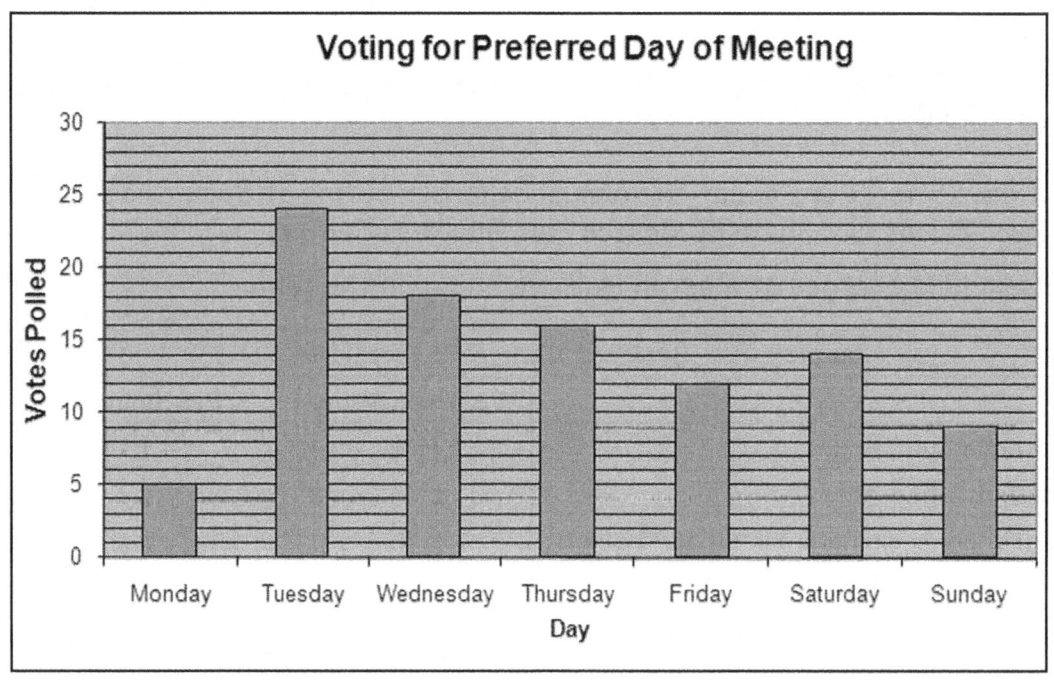

21 Which is the **closest** fraction of the voting members who prefer Tuesday as the monthly meeting day?

    A $\frac{1}{4}$

    B $\frac{1}{5}$

    C $\frac{1}{6}$

    D $\frac{1}{7}$

Questions 22 to 25 are about wedding guests travelling by ferry to the Isle of Wight.

A group of twelve adults plans to travel to the Isle of Wight by passenger ferry to attend a wedding. They intend to travel on Friday afternoon and return on Sunday evening. They book tickets for return ferry passage from Portsmouth to Ryde.

The fares are shown in the table:

Isle of Wight Passenger Ferry: Portsmouth to Ryde, per person		
Day Return	One way	Return
£35	£19	£38
Discount of 10% for groups of 10 or more on return bookings		

22   The total cost for their return journey is:

A   £456.00
B   £410.40
C   £380.00
D   £350.00

The wedding guests need to arrive in Ryde before 6:15pm and check the timetable:

Ferry Number	Depart Portsmouth	Arrive Ryde
PRO102	0820	0935
PR1205	1050	1205
PR1615	1320	1435
PR2050	1550	1705
PR2152	1820	1935
PR1111	2050	2205

23   What is the Ferry Number that represents the **latest** ferry that they could catch?

A   PR1111
B   PR1615
C   PR2050
D   PR2152

Please go on to the next page

Another group of wedding guests travel to the same wedding from Fishbourne by car ferry to attend the wedding to be held at 12:30 on Saturday 2nd June.

The car ferry timetable is shown here:

Redsail Car Ferry Timetable Fishbourne to Ryde Approximate sailing time: 1 hour 35 minutes		
		Departure Time
Daily crossing	5th May to 26th September	07:30 & 17:45
Weekend crossings	10th May to 15th June	09:25
	21st June to 1st October	10:35

24 On the day of the wedding, at what approximate time is the car ferry due to arrive in Ryde?

    A   09:25
    B   10:35
    C   11:00
    D   12:00

25 Ten of the wedding guests stay for two nights at Shingle Bay Motel. Wedding parties of ten or more receive a discount of 15%. The total bill before the discount is £590. What is the cost per guest at the discounted rate?

    A   £50.15
    B   £51.50
    C   £55.10
    D   £59.15

26 One tier of wedding cake is cut into slices for guests to take home as a souvenir. The cake tier measures 10cm x 30cm x 30cm.

Slices of cake measure 10cm x 5cm x 2cm. How many slices can be cut from the cake?

    A   100
    B   90
    C   70
    D   60

Please go on to the next page

Questions 27 to 29 are about plans for a new music venue.

27 An architectural artist in a council's planning department makes a scale model of a planned music venue to be built in a new arts and leisure complex.

The model is constructed to the scale 1 : 50.

The length of the music venue is to be 45 metres.

What is the length of the scale model?

A   90mm
B   225mm
C   900mm
D   2250mm

The plan shows the symmetrical seating area in the music venue.

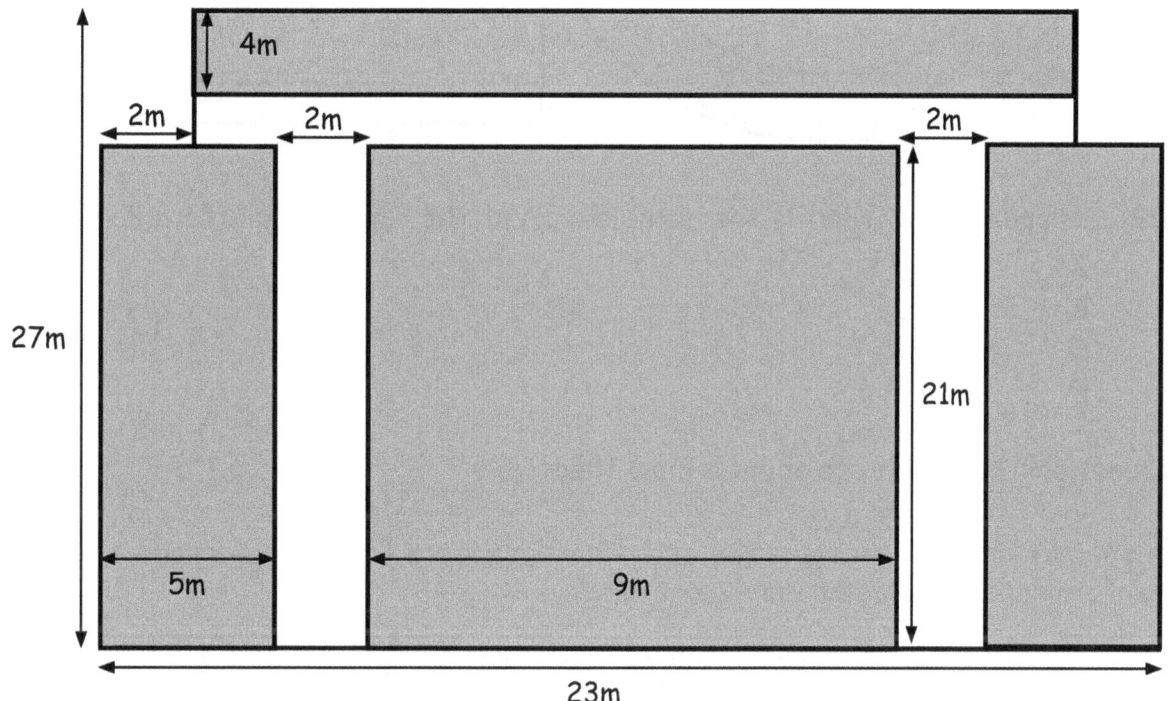

Diagram not to scale

28 The shaded blocks will have seats installed and each seat takes up 100cm x 100cm. What is the total number of seats that can be installed in the seating area?

A   345
B   400
C   450
D   475

The venue stage will have a revolving platform with a radius of 2.95 metres.

For Health and Safety reasons, no equipment can be placed within 0.5 metres of the perimeter of the revolving platform.

Diagram not to scale

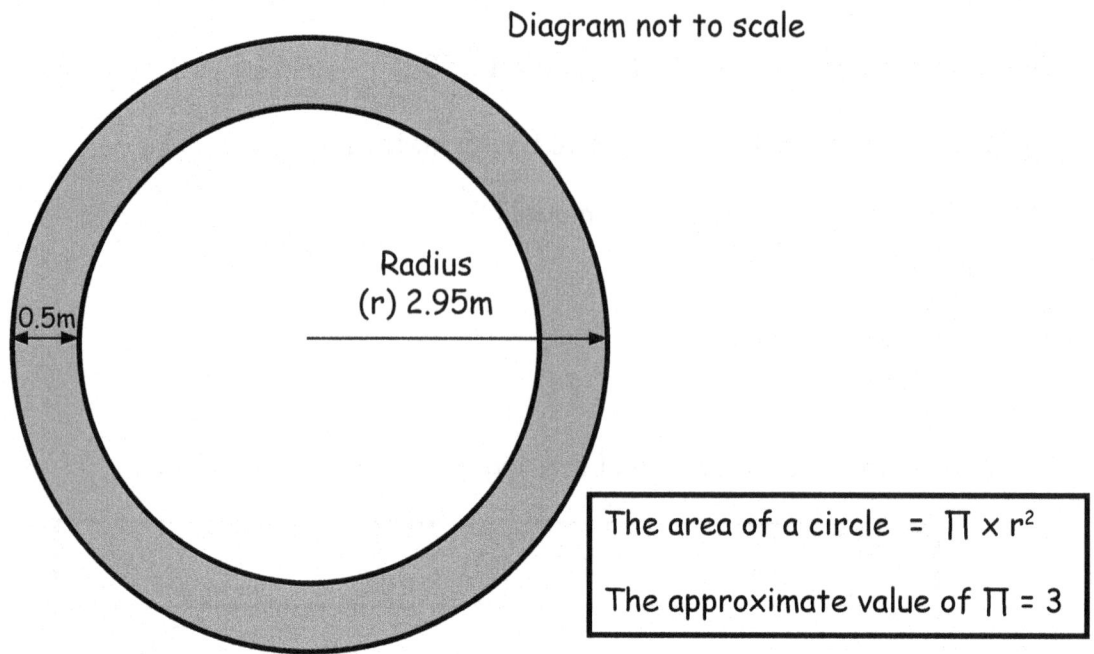

The area of a circle = $\pi \times r^2$

The approximate value of π = 3

29   Approximately what is the useable area of the revolving platform?

A   18.0m²
B   26.1m²
C   27.0m²
D   30.0m²

Questions 30 and 31 are about a local theatre.

A theatre manager analyses the attendance numbers for the afternoon performances of Aladdin during one week.

Day	Mon	Tue	Wed	Thur	Fri	Sat	Sun
Adults	103	67	115	141	192	210	228
Children	63	52	40	38	82	60	85

30   What is the mean number of children attending the afternoon performances during the week?

A   84
B   70
C   60
D   52

31 Tickets for the afternoon performance cost £7.00 for adults and £3.50 for children. What is the total income from ticket sales for the afternoon performance on Sunday?

A   £1596.00
B   £1876.50
C   £1893.50
D   £3286.50

Questions 32 to 34 are about ladies who play golf at Lizard Lane Links Club.

In a round of golf, Moira and Jean walk three-and-a-quarter miles.

1 mile = 1.6 kilometres

32 How far do they walk in kilometres?

A   4.8km
B   5.2km
C   5.4km
D   5.6km

Moira's scores for this playing season are shown in the frequency table:

Score	68	69	70	71	72	73	74	75	76	77	78
Frequency	1	4	2	2	2	3	2	2	1	2	1

33 Moira's ranking at Lizard Lane Links Club is calculated from her median score. What is Moira's median score?

A   71.0
B   71.5
C   72.0
D   72.5

34 Prizes are awarded each season to players with scores of less than 74 for at least 50% of playing season.

What is the approximate percentage of Moira's scores that are less than 74?

A   65%
B   64%
C   56%
D   44%

Please go on to the next page

Questions 35 and 36 are about a hospital physiotherapy department.

A physiotherapist is asked to analyse the ages of patients attending appointments over three days. She draws up a table and gathers this data:

Physiotherapy Dept.	Wednesday	Thursday	Friday
Females	42, 76, 91, 19, 17, 52	15, 26, 81, 44 12, 58, 74,	53, 38, 14, 58, 81, 64
Males	68, 77, 22, 34, 26, 18, 72	44, 61, 78, 36 24, 18, 13	88, 38, 92, 14, 86, 12

The data is used to construct this frequency table:

Frequency (years of age)	19 and under	20 to 39	40 to 59	60 and over
Females	5	2	6	6
Males	6	6	1	8

35 An error has occurred in transferring the data to the frequency table. Which is the incorrect entry?

- **A** males 19 and under
- **B** males 20 to 39
- **C** males 40 to 59
- **D** males 60 and over

Please go on to the next page

36 A notice on the wall in the waiting room contains a graph that shows the percentage of patients who fail to turn up for appointments.

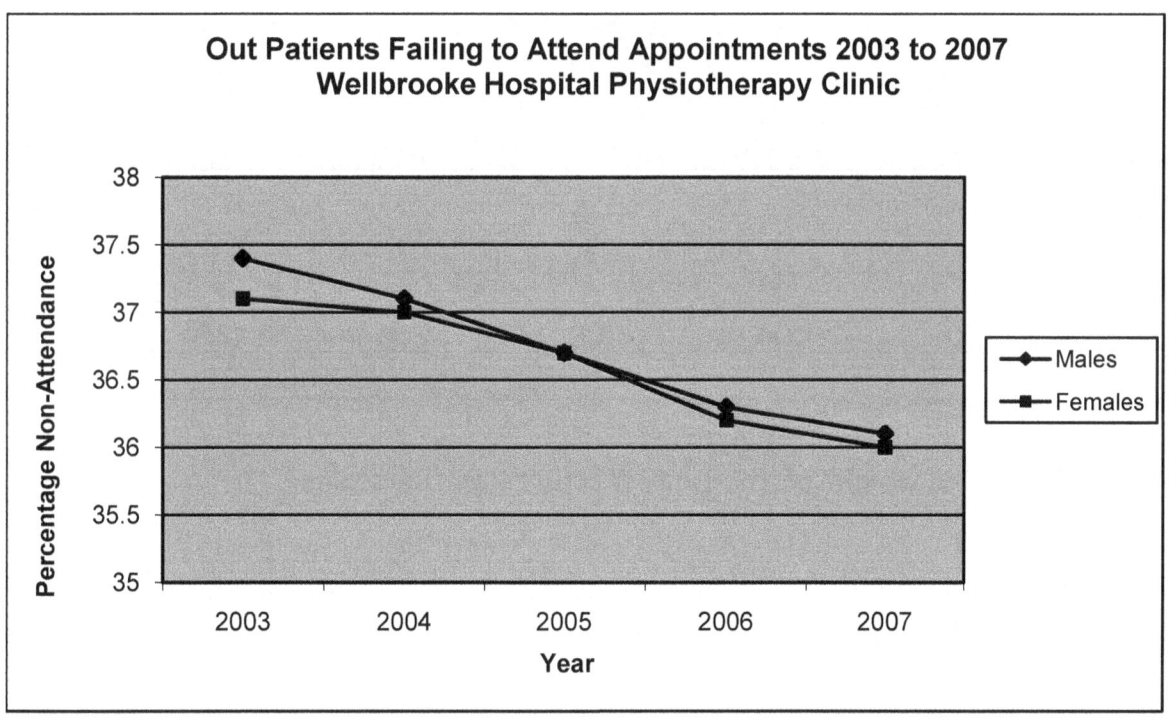

What trend does the line graph show?

A   an increase in patients failing to attend appointments
B   more females failing to attend appointments than males
C   an increase in males failing to attend appointments
D   a decrease in both males and females failing to attend appointments

Questions 37 to 40 are about Anja's mobile candy stall.

A candy stall holder fills clear plastic boxes with pineapple chunks.

Each chunk measures 1cm x 1cm x 1cm.

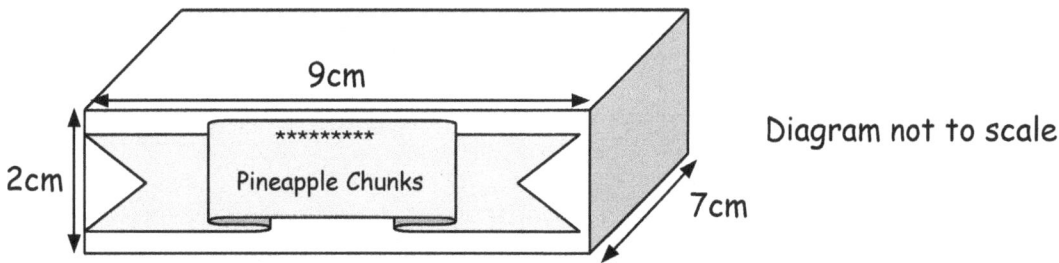

Diagram not to scale

37 How many chunks fit into each box?

A   144
B   140
C   126
D   120

38 A customer selects a scoop of Dolly Mixtures from the stall's 'Pick-and-Mix' sweets and puts them on the scales to weigh.

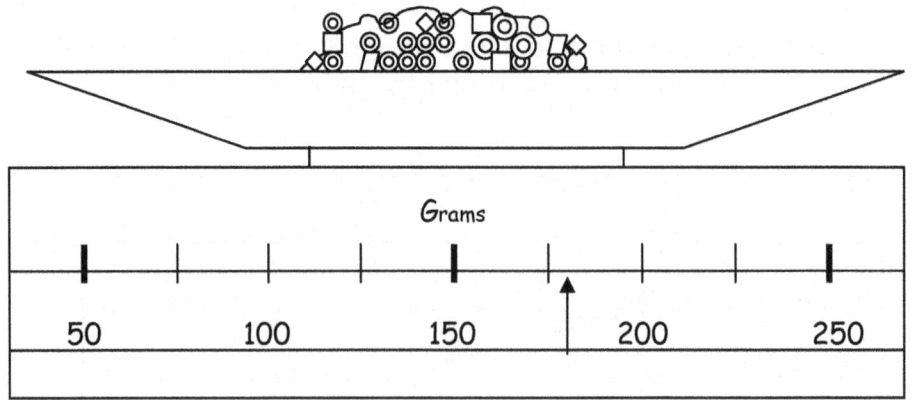

What is the weight of the Dolly Mixtures on the scales?

A 155g
B 160g
C 175g
D 180g

39 Anja travels from her home to her pitch in the shopping mall five days a week. The fuel tank in her van holds 34 litres of diesel when full.

Fuel gauge Monday morning

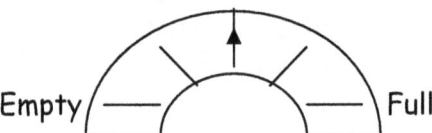

Fuel gauge Saturday evening

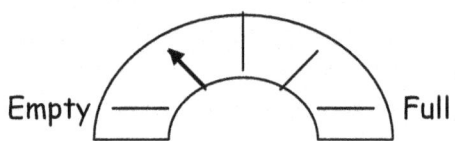

Approximately how many litres of diesel fuel does Anja use each day to travel from home to the shopping mall and back?

A 1.6 litres
B 1.7 litres
C 1.8 litres
D 1.9 litres

Please go on to the next page

40 Anja's mobile candy stall is part of a franchise company linked to a confectionery manufacturer. Four other franchise stalls operate in the North West of England.

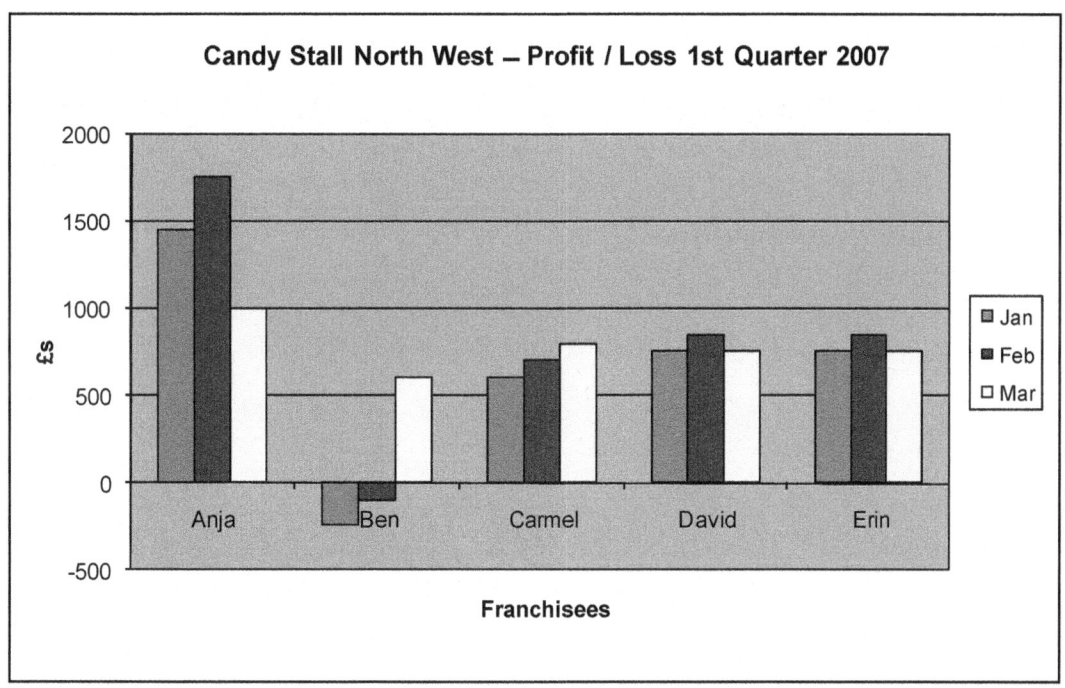

The chart shows profit/loss for the five franchise candy stalls over three months.

Which of the following statements is **false**?

A   Ben started the year with a loss but is in profit by March.
B   Every franchisee was in profit in March.
C   Carmel was the most profitable of all the franchisees in March.
D   David and Erin made identical profits each month.

End of Paper

Practice Multiple-choice Paper
Suitable for:

Key Skills Level 2 Application of Number
Level 2 Adult Numeracy

Paper Six

YOU NEED

- This test paper.
- A pen.
- A pencil and eraser.
- An Answer Sheet.
- A ruler marked in centimetres and millimetres

You may NOT use a calculator.
You may use a bilingual dictionary.
There are 40 questions on this paper. Try to answer ALL the questions.
When you have completed the questions you must check your answers, then check them again.

YOU HAVE QUARTER OF AN HOUR TO READ THE PAPER
AND ONE HOUR TO COMPLETE THE 40 QUESTIONS

INSTRUCTIONS

- Make sure you write your name and today's date on the Answer Sheet. Use a pen to do this.
- Use a pencil to mark your answers so if you change your mind you can erase your choice and select another.
- Make sure that for each question you have only selected one answer. If you select more than one, the answer will not be marked.
- Read each question carefully before you select an answer.

Note for learners and tutors: This is a practice test that has been designed to closely resemble the questions and question styles of a "live" paper.

Questions 1 to 7 are about the planning of a camping holiday in Northern France.

The table shows web site information on camping package holidays. Prices are in £s for a family group of up to six persons including the cost of the Cherbourg car ferry.

Town	Pont L'Eveque		Houlgate		St Pair-Sur-Mer	
Camping Site	Château Le Brevedent		La Vallée		Château Les Eaux	
Number of Days:	7	14	7	14	7	14
Departure Days:	Sat	Sun	Tues	Fri	Mon	Thurs
Pricing Period						
26 May – 16 Jun	335	415	329	420	350	425
17 Jun – 30 Jun	345	425	339	430	360	435
1 Jul – 21 Jul	355	435	349	440	370	445
22 Jul – 4 Aug	365	445	359	450	380	455
5 Aug – 18 Aug	385	465	380	475	400	470
19 Aug – 30 Aug	400	480	400	500	430	505
31 Aug – 22 Sep	380	440	360	420	360	425

Calendar	July				
Monday		7	14	21	28
Tuesday	1	8	15	22	29
Wednesday	2	9	16	23	30
Thursday	3	10	17	24	31
Friday	4	11	18	25	
Saturday	5	12	19	26	
Sunday	6	13	20	27	

1   What is the earliest departure date that the family can book for a 14-day camping holiday at Château Le Brevedent in Pont L'Eveque in July?

    A   1st July
    B   2nd July
    C   6th July
    D   7th July

2   Using web site information on camping package holidays, the mean average price of 7-day camping holidays at La Vallée in Houlgate is £359.

    To the nearest pound, how much more is the mean price for 7-day camping holidays at the Château Les Eaux?

    A   £7
    B   £11
    C   £15
    D   £20

The line graph shows exchange rates between Euros and British pounds.

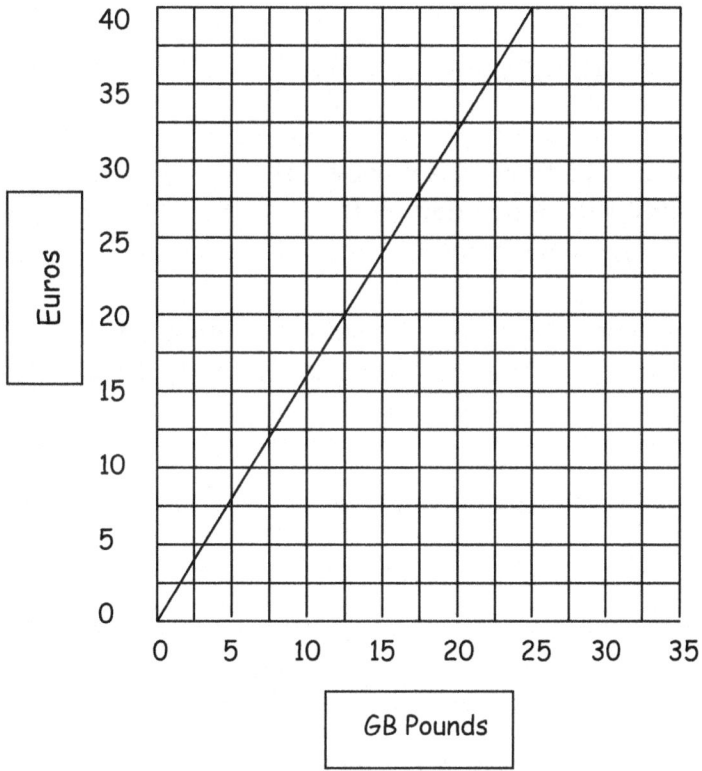

3 Mum decides she needs at least 60 Euros in cash on entry to France. How much is this amount in British pounds?

A   £37.50
B   £40.00
C   £50.50
D   £96.00

The distance between Cherbourg and Pont L'Eveque is 97 miles.

> **1 mile = 1.6 kilometres**

4 To the nearest kilometre, what is the distance between Cherbourg and Pont L'Eveque?

A   61km
B   122km
C   146km
D   155km

5 On a tour guide map of Northern France the distance between Pont L'Eveque and Paris is 6.9cm. The scale of the map is 1cm : 20km.

What is the real distance between Pont L'Eveque and Paris?

A   140km
B   138km
C   35km
D   13.8km

6   The temperature gauge in the family's car measures in degrees Fahrenheit (°F) but not in degrees Celsius (°C).

To convert °F to °C, follow these steps:

- Subtract 32 from the number of °F
- Multiply the remainder by 5
- Divide the answer by 9.

On arrival in France the temperature gauge showed 80°F. What was the temperature, to the nearest whole degree, in degrees Celsius?

A   26°C
B   27°C
C   44°C
D   45°C

7   The distance between Pont L'Eveque and St. Malo is 93 kilometres .

The car journey takes 1 hour and 30 minutes.

Average speed = distance / time

The average speed of the car, to the nearest kilometre per hour, during the journey is ...?

A   32km
B   52km
C   62km
D   72km

Please go on to the next page

Questions 8 to 11 are about landscape gardening activities.

A householder wants to replace the old lawn with fresh garden turf.

This is a sketch of the lawn dimensions:

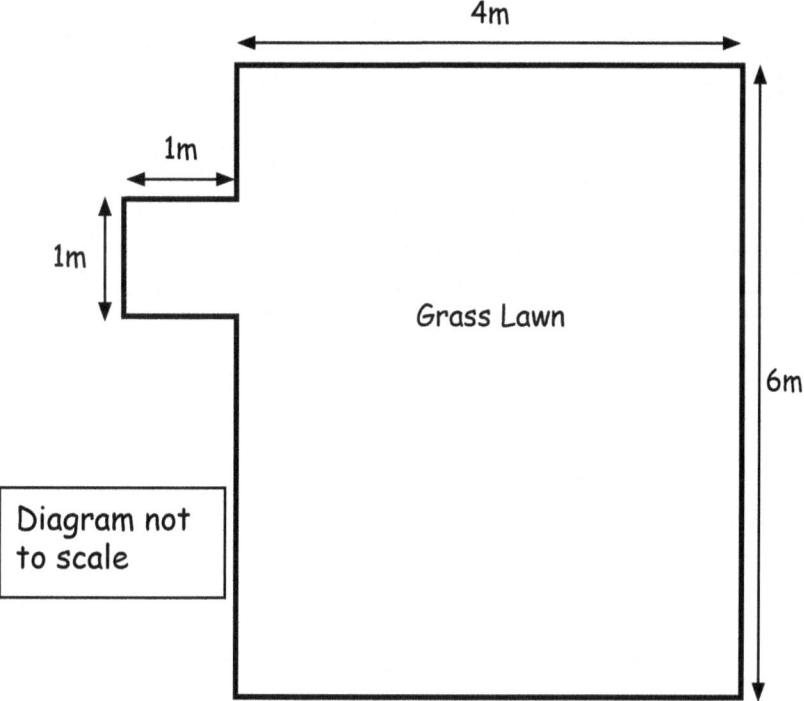

8   New turf costs £4.21 per square metre. How much does it cost the householder to re-turf the grass lawn?

   A   £109.46
   B   £105.25
   C   £101.04
   D   £50.52

9   A landscape gardener charges a fixed call-out fee of £49.95 plus £9.95 per hour. He estimates that it will take 6 hours 45 minutes to remove the old lawn and replace it with fresh turf. His quote for the job is £117.11. Which calculation finds the closest approximation to the quoted price for the job?

   A   (10 x 6) + 50
   B   (10 + 50) x 6
   C   (10 + 50) x 7
   D   (10 x 7) + 50

Please go on to the next page

10 A conservatory is to be built in the garden. A rectangular concrete base is required measuring 2.5m by 3m. The local builder charges £110 to lay 5m² of concrete.

What is the cost of laying the concrete base for the conservatory?

A £220
B £165
C £135
D £125

The builder's method of mixing the ingredients of the concrete is as follows:

> **Combine sand, gravel and wet cement in the ratio 3 : 2 : 1**

11 If the builder uses 48 shovels of gravel in the concrete mix, how many shovels of sand are needed to keep the ratio correct?

A 72
B 54
C 36
D 24

Questions 12 and 13 are about temperatures in the continents around the world.

Liam compares temperatures around the world for his geography homework. He draws a table showing the Lowest and Highest recorded temperatures in five continents.

Continent	Lowest Recorded Temperatures °C	Highest Recorded Temperatures °C
Africa	-25	58
America	-54	54
Asia	-57	55
Australia	-16	54
Europe	-52	49

12 What is the mean of the Lowest Recorded Temperatures?

A -36°C
B -36.8°C
C -40.8°C
D -50.9°C

13 Which of these figures is the median of the Highest Recorded Temperatures?

A 49°C
B 54°C
C 55°C
D 58°C

Questions 14 to 16 are about commercial airfreight cargo transport.

14 The temperature inside a cargo aircraft is 62°F. Mid-flight the temperature outside the aeroplane is -66°F. The temperature difference between the inside and outside of the aeroplane is

A   -4°F
B   4°F
C   124°F
D   128°F

15 A transport aeroplane weighs 220 tonnes when empty. The ratio of the maximum weight on take-off to the weight of the empty aeroplane is 2.4 : 1.

What is the maximum weight of cargo the aeroplane can carry?

A   240 tonnes
B   242 tonnes
C   484 tonnes
D   528 tonnes

16 Diesel fuelled trucks move cargo on the airfield. A full fuel tank holds 40 litres.

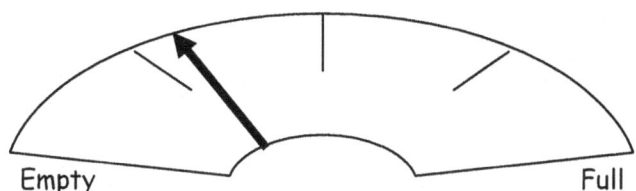

What is the nearest estimate of the fuel left in this tank?

A   11 litres
B   13 litres
C   15 litres
D   20 litres

Questions 17 and 18 are about a dating agency.

17 A dating agency currently has 2487 members on its database. 1003 of the agency's clients are men. What is the closest approximation of the ratio of **women to men**?

A   12 : 5
B   5 : 2
C   3 : 2
D   7 : 5

18 The dating agency database manager conducted a survey of current members. The result was that 5/8 of the male users had dated at least twice in the past month. What is the fraction 5/8 as a percentage?

    A   40%
    B   58%
    C   62.5%
    D   65%

Questions 19 and 20 are about a beauty therapy salon and spa.

19 An attendant of the spa pool in a beauty therapy salon checks and records the pH level six times in a day.

Time of Day	9:00am	10:00am	11:00am	1:00pm	3:00pm	5:00pm
pH level	8.2	7.4	7.3	7.9	7.1	7.6

What is the range of the pH levels in the spa pool?

    A   1.9
    B   1.1
    C   0.8
    D   0.5

20 The spa pool is filled to a depth of 1.25m. The diagram shows the plan view of the spa pool with length and width measurements indicated.

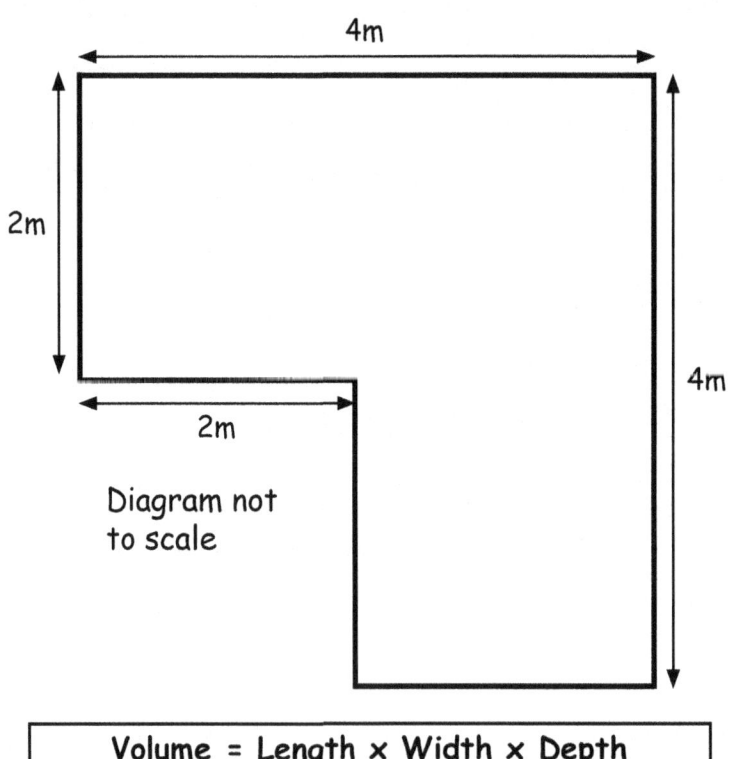

Diagram not to scale

Volume = Length x Width x Depth

What is the volume of water, in cubic metres, required to fill the spa pool?

A   15m³
B   20m³
C   22m³
D   24m³

Questions 21 to 25 are about a bus company.

21   Elland Bus company employs a total of 590 people. 147 of the employees work part time hours. Approximately what fraction of the employees work full time hours?

A $\frac{3}{6}$

B $\frac{3}{5}$

C $\frac{3}{4}$

D $\frac{2}{3}$

22   The table shows the profit/loss status of two bus routes over a 5-year period.

Year	2003	2004	2005	2006	2007
Route 101	20500	35000	40000	56000	65000
Route 129	-25000	-20000	10000	60000	75000

A chart is made from the data in the table.

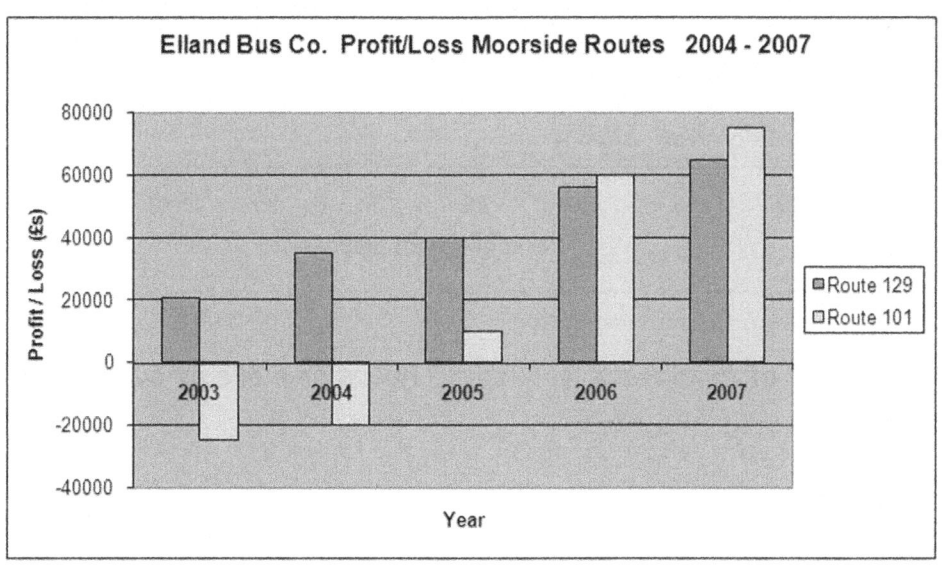

The chart is not accurate. What is the error?

A   some data are incorrectly plotted
B   the range of the vertical axis is too small
C   the scale of the horizontal axis is incorrect
D   the Route labels in the key are in the wrong order

23 The weekly earnings of Elland Bus Co employees are shown in the table.

Employee Numbers	Weekly Earnings
147	Under £200
325	£200 - £299
82	£300 - £399
24	£400 - £499
12	£500 and Above

What percentage of the company's employees earn £300 or more per week?

A   10%
B   15%
C   20%
D   25%

24 The chart shows the distribution of overtime hours worked by drivers over a 6-month period from July to December.

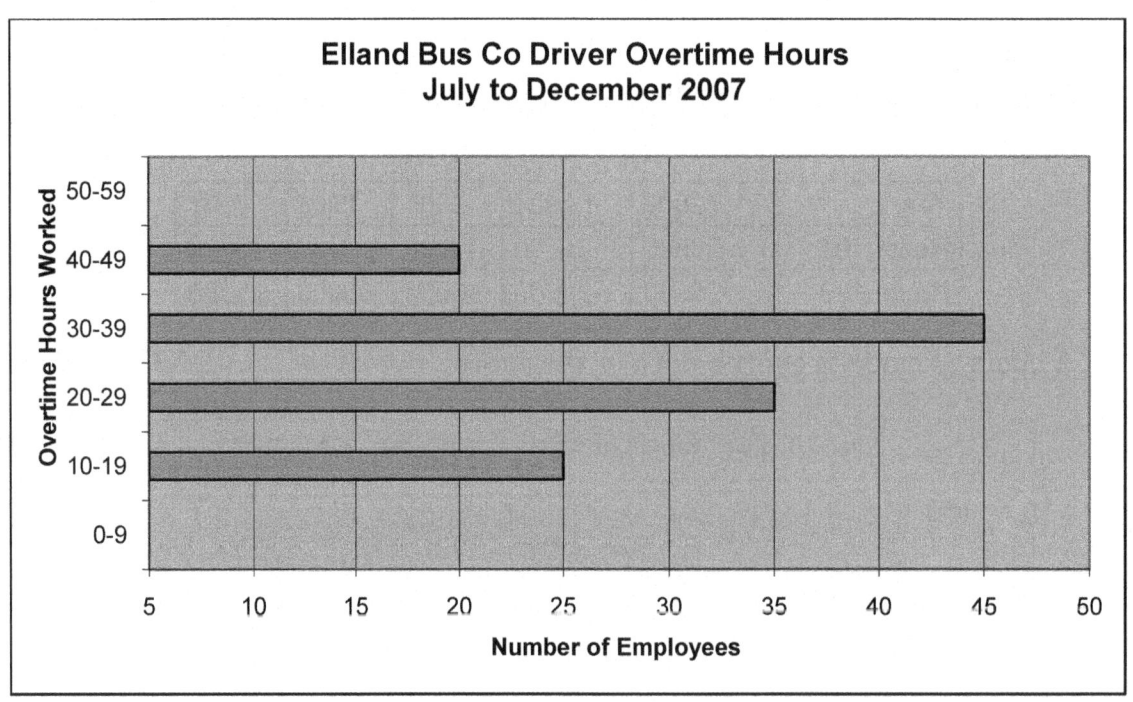

What number of employees worked 30 hours or more of overtime?

A   70
B   65
C   45
D   25

Please go on to the next page

25 This is the timetable for the express X190 bus service.

Elland – Maston X190 Monday to Friday											
Elland	05.39	06.12	07.11	08.23	08.47	09.26	09.56		16.56	17.26	17.56
Gateshead	05.48	06.21	07.20	08.33	08.57	09.36	—		17.06	17.40	18.06
Tunworth	05.53	06.26	07.25	08.38	09.02	09.41	10.10		17.12	17.45	18.12
Relham	05.59	06.31	07.30	08.43	09.07	—	10.15	then these minutes past each hour until	17.17	17.50	18.17
Portwick	06.04	06.37	07.36	08.49	09.13	09.49	10.20		17.23	17.55	18.23
Hartburn	06.08	06.40	07.39	08.53	09.16	09.52	10.23		17.26	17.58	18.26
Maston	06.12	06.45	07.44	08.57	09.20	09.57	10.28		17.31	18.03	18.31
Maston – Elland X190 Monday to Friday											
Maston	06.35	07.15	07.31	07.44	08.09	09.10	09.41		17.17	17.47	18.19
Hartburn	06.40	07.20	07.36	07.49	08.14	09.15	09.46		17.19	17.52	18.24
Portwick	06.43	07.23	07.39	07.52	08.17	09.18	09.49		17.22	17.55	18.27
Relham	06.48	—	07.43	07.57	08.22	09.23	09.54		17.27	18.00	18.32
Tunworth	06.52	—	07.47	08.01	08.26	09.27	09.58		17.31	18.04	18.36
Gateshead	06.56	—	07.51	08.06	08.30	09.31	—		17.35	—	18.40
Elland	07.04	07.50	08.01	08.14	08.42	09.43	10.13		17.44	18.16	18.48

How many buses depart from Elland between 10am and 4pm?

A   6
B   10
C   12
D   14

Please go on to the next page

Questions 26 to 29 are about visitor numbers at leisure attractions.

26   This table records the number of visitors to Althorpe Tower theme park, in the past ten years.

Year	Visitor Numbers (millions)
1998	1.43
1999	1.31
2000	1.51
2001	1.50
2002	1.80
2003	1.49
2004	1.87
2005	1.99
2006	2.01
2007	2.09

Which of these calculations correctly works out the percentage increase in visitor numbers between the years 1998 and 2007?

A $\dfrac{2.09 - 1.43}{2.09} \times 100$

B $\dfrac{2.09 - 1.43}{1.43} \times 100$

C $\dfrac{1.43}{2.09} \times 100$

D $\dfrac{2.09}{1.43} \times 100$

27   The table shows the number of visitors who buy a Day Pass to use the leisure facilities at Horton Fun Park during a Bank Holiday Weekend.

Day Pass Cost per Person	Saturday	Sunday	Bank Holiday Monday	Total
Adult £4.50	194	206	200	600
Child £3.00	80	70	85	235
Total:	274	276	285	835

What is the total income from sales of Day Passes at Horton Fun Park over the 3-day Bank Holiday Weekend?

A   £3405
B   £2700
C   £2400
D   £1800

28 The graph displays Horton Fun Park's visitor numbers between 2001 and 2007.

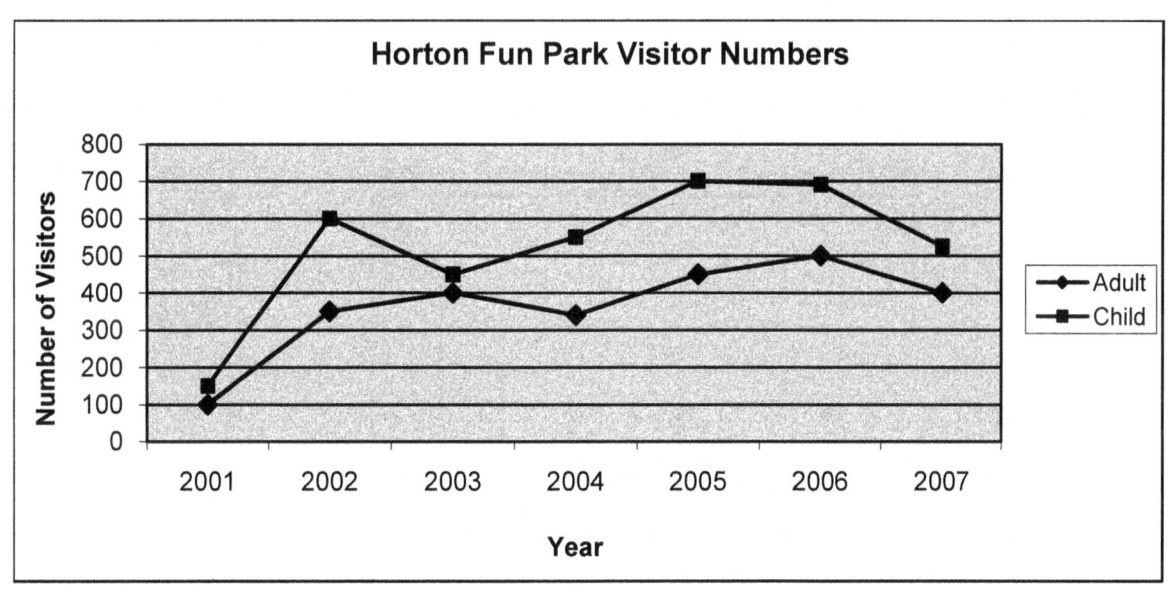

What was the total number of visitors, to the nearest 50, who visited Horton Fun Park in 2004?

A   600
B   850
C   900
D   1150

29 A coach driver brings in holidaymakers from a seaside resort to visit Compton Hall Country Theme Park.

The table shows the coach driver's time sheet for 4 weeks in May:

Week	Thursday	Friday	Saturday	Sunday	Monday
1	7	7	8	6	9
2	8	7	8	6	7
3	7	7	8	6	7
4	7	7	7	6	9

The coach driver is paid: £9.50 per hour plus £13 per day for expenses.

Which calculation works out the driver's total pay and expenses for week 3?

A   (5 × £9.50) + (5 × £13)
B   (35 × £9.50) + 13
C   (£9.50 + 13) × 35
D   (35 × £9.50) + (5 × £13)

Questions 30 and 31 are about gravy granules.

30 Cartons of gravy granules are packed into a box for delivery to a supermarket. The diagram shows the measurements of a carton and the packing box.

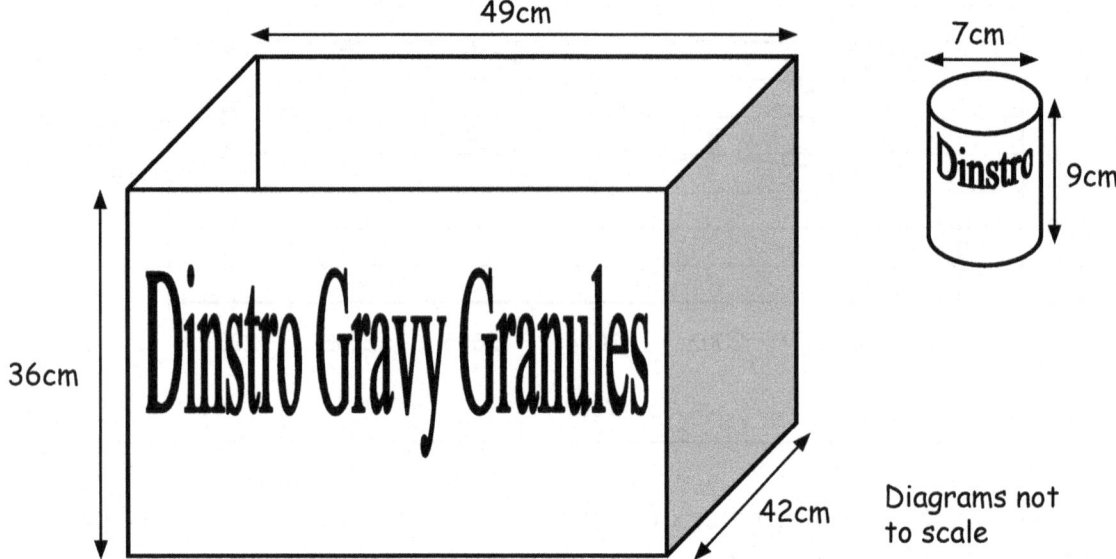

What is the total number of cartons that can fit into the packing box?

A   70
B   140
C   168
D   196

31 There are 115 calories in one ounce of gravy granules. One ounce is approximately equivalent to 30 grams.

How many calories are there in a 300g carton of gravy granules?

A   1250 calories
B   1150 calories
C   660 calories
D   330 calories

Please go on to the next page

Questions 32 to 34 are about the graph representing a patient's body temperature.

The graph records the rise and fall of a patient's temperature in the recovery room after a surgical operation which ended at 8 o'clock in the evening.

Temperature Chart

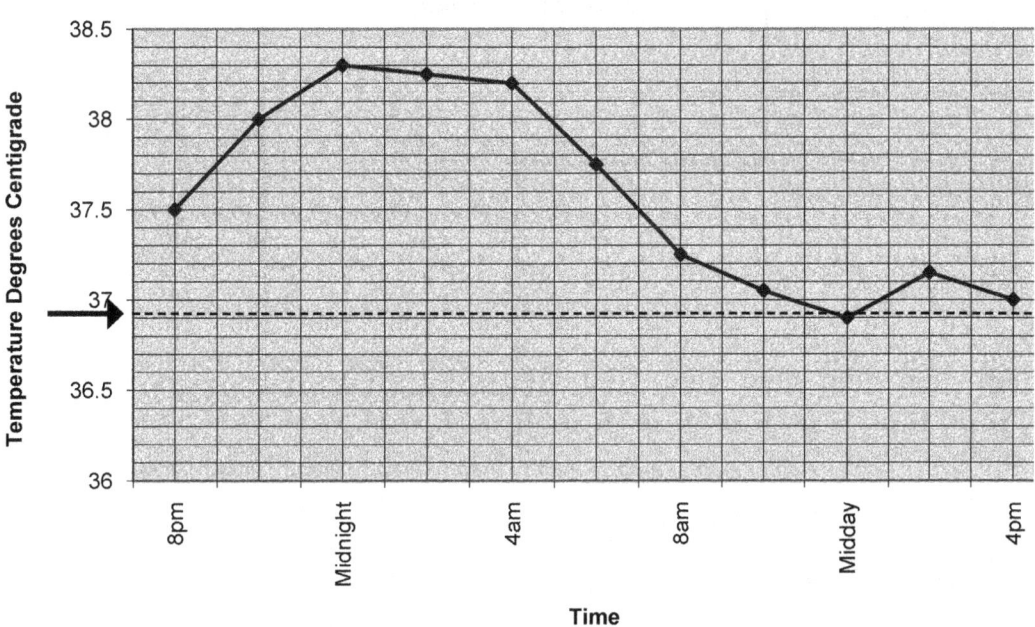

32   What was the patient's temperature at 6 o'clock in the morning?

A   38.30°C
B   38.25°C
C   37.75°C
D   37.50°C

33   How many hours did it take for the patient's temperature to return to normal from the end of the operation? (⟶ arrow indicates normal temperature)

A   16 hours
B   14 hours
C   12 hours
D   8 hours

34   What is the approximate range of the patient's temperature in the graph?

A   2.5°C
B   1.5°C
C   1.4°C
D   0.4°C

Questions 35 to 39 are about plans for a shop refurbishment.

35   A shop owner draws a scale plan of an extension to his premises.

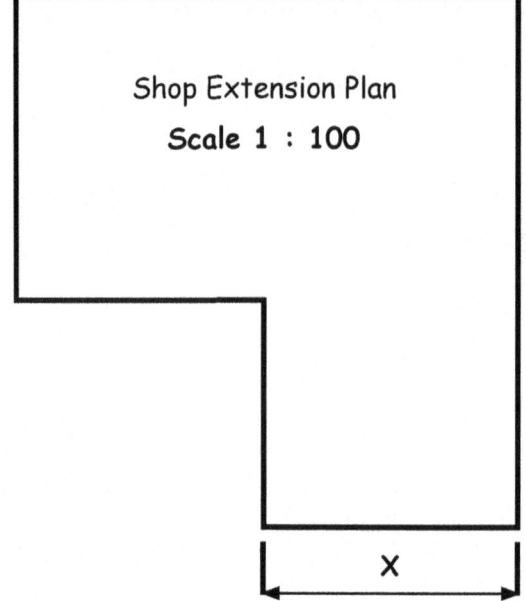

Shop Extension Plan
Scale 1 : 100

What will be the width of the actual extension at X?

A   35mm
B   350mm
C   3500mm
D   35000mm

36   A scale drawing of one of the new shop windows is shown in the diagram but the scale of the drawing has been omitted. The size of the actual window is 2.4m high and 4.3m wide.

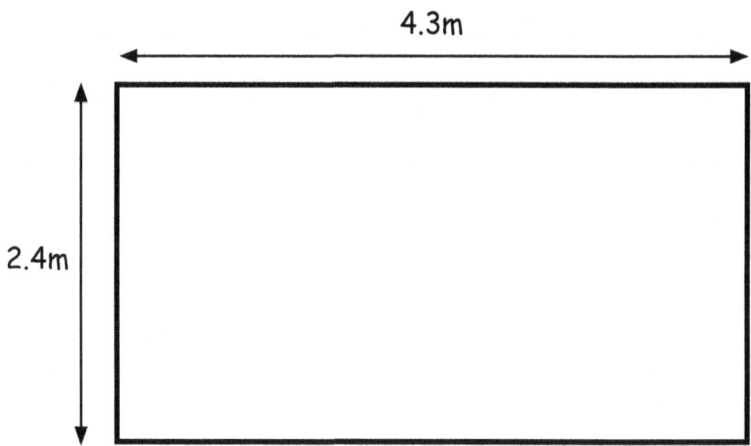

What is the scale of the drawing?

A   1 : 100
B   1 : 50
C   1 : 25
D   1 : 10

37 The floor of the extension is made from self-levelling concrete. The ready-mixed dry material for the concrete requires 2.5 litres of water to every 20 kilograms of dry concrete mix. To the nearest half-litre what is the amount of water to be added to 140 kilograms of dry concrete mix?

A 17.5 litres
B 15.25 litres
C 14.0 litres
D 12.5 litres

38 The stockroom floor is to be re-tiled as part of the shop refurbishment. The floor dimensions are show in the diagram.

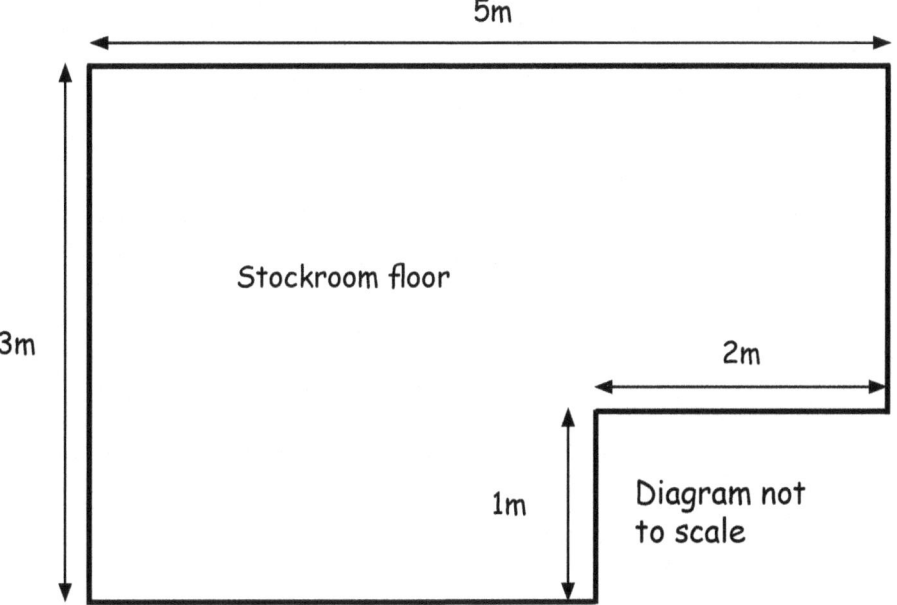

Ceramic floor tiles measure 0.25m x 0.25m. How many tiles are needed to cover this stock room floor?

A 104
B 120
C 192
D 208

39 The gas bill for the shop for April 1st to June 30th includes a fixed standing charge of £58.35 and each gas unit used costs 9.5 pence.

Which calculating method should be used to find the total gas bill, in pounds, for 1423 units?

A $\dfrac{9.5 \times 100}{100} + 58.35$

B $9.5 \times 1423 + 58.35$

C $\dfrac{9.5 \times 1423}{100} + 58.35$

D $(58.35 + 9.5) \times 1423$

40 Dad wants to oven-roast a 4-kilogram turkey. To ensure that the turkey is safe to eat he must allow 30 minutes cooking time per kilogram plus an extra 35 minutes. He calculates that the turkey will need to cook for 155 minutes.

Which calculation can he use to check if this is the correct cooking time?

A $\frac{155}{30} - 35$

B $\frac{155 - 35}{30}$

C $\frac{155}{35} - 30$

D $30 \times 35 \div 155$

End of Paper

Practice Multiple-choice Paper
Suitable for:

Key Skills Level 2 Application of Number
Level 2 Adult Numeracy

Paper Seven

YOU NEED

- This test paper.
- A pen.
- A pencil and eraser.
- An Answer Sheet.
- A ruler marked in centimetres and millimetres

You may NOT use a calculator.
You may use a bilingual dictionary.
There are 40 questions on this paper. Try to answer ALL the questions.
When you have completed the questions you must check your answers, then check them again.

YOU HAVE QUARTER OF AN HOUR TO READ THE PAPER
AND ONE HOUR TO COMPLETE THE 40 QUESTIONS

INSTRUCTIONS

- Make sure you write your name and today's date on the Answer Sheet. Use a pen to do this.
- Use a pencil to mark your answers so if you change your mind you can erase your choice and select another.
- Make sure that for each question you have only selected one answer. If you select more than one, the answer will not be marked.
- Read each question carefully before you select an answer.

Note for learners and tutors: This is a practice test that has been designed to closely resemble the questions and question styles of a "live" paper.

Questions 1 to 3 are about employees' journey time to the workplace.

The Allied Building Society carried out a survey of the travelling times taken by employees to get to work. The table shows the survey results from the Harton office:

Allied Building Society (Harton Office) Employee Journey Time To Work (minutes)									
15	25	45	15	15	20	30	35	35	25
20	10	30	35	30	45	25	10	40	55

1 What fraction of the employees in the Harton Office have a journey time of 20 minutes or less from home to the workplace?

A $\dfrac{1}{4}$

B $\dfrac{7}{20}$

C $\dfrac{1}{2}$

D $\dfrac{3}{4}$

2 What is the range of journey times for employees at the Harton Branch?

A 10 minutes
B 20 minutes
C 45 minutes
D 55 minutes

3 What is the mean travelling time, in minutes?

A 27 minutes
B 28 minutes
C 29 minutes
D 30 minutes

Please go on to the next page

Questions 4 and 5 are about the number of mobile phones in households in Castle Town.

A market researcher analyses the responses, from a survey carried out on one day, in a frequency chart:

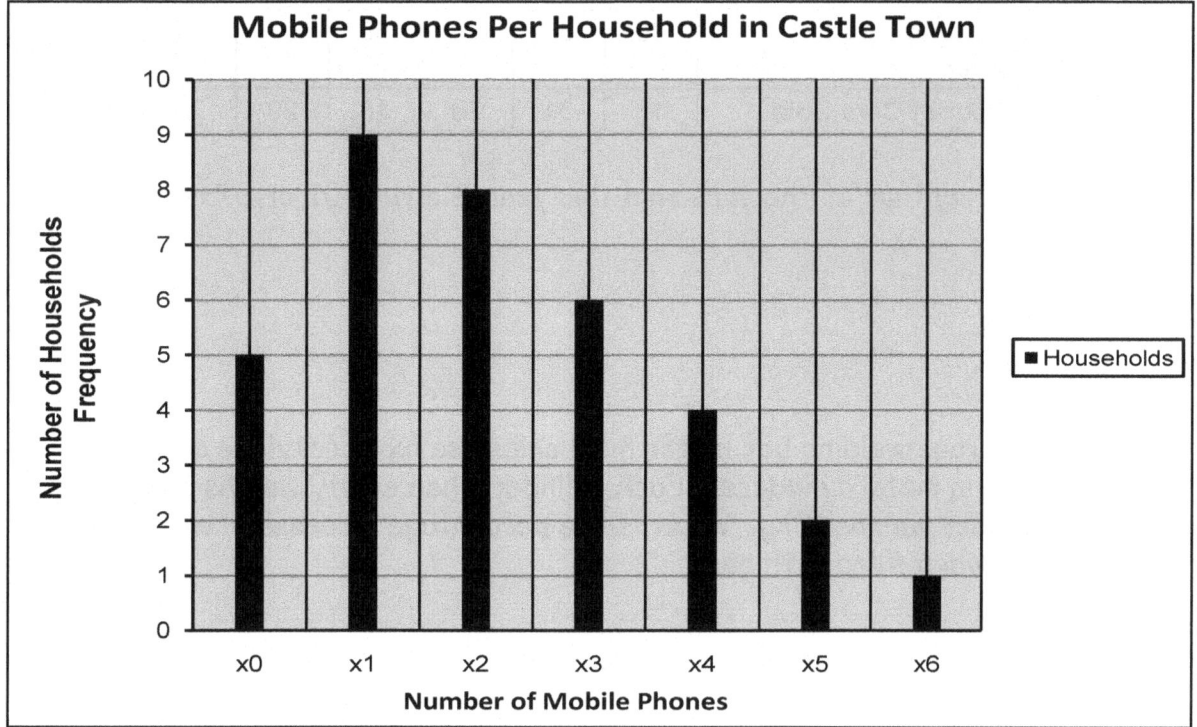

4   How many households had at least two mobile phones, but fewer than 5?

A   18
B   19
C   21
D   27

5   What is the mode of the number of mobile phones in Castle Town households?

A   x5
B   x3
C   x2
D   x1

Questions 6 to 8 are about a car dealership garage.

A car salesman records his sales for this year. He creates a table with the number of cars sold against the registration code number for the year of manufacture.

Registration Code Number	05	55	06	56	07	57
Number of Cars Sold	15	31	26	18	29	31

6   What percentage of the cars sold this year is either 07 or 57 registration?

A   20%
B   33%
C   40%
D   66%

7   In the garage welding bay motor mechanics use oxy-acetylene gas which is stored in metal cylinders. Each cylinder when empty weighs 15kg. A full cylinder weighs 18kg. What is the percentage increase in weight of a cylinder when filled with gas?

A   20%
B   18%
C   15%
D   3%

8   Motor mechanics change the lubricating oil in car engines during routine servicing. The table shows grades of oil which are recommended for use in different temperature ranges.

Class of Oil	Temperature Range (°C)
SLO 20W-45	-20 to 45
SLO 15W-45	-15 to 45
SLO 15W-40	-15 to 40
SLO 10W-40	-10 to 40

Which Class of Oil in the table has the greatest temperature range?

A   SLO 20W-45
B   SLO 15W-45
C   SLO 15W-40
D   SLO 10W-40

Please go on to the next page

Questions 9 and 10 are about preparations for a wedding.

9 Dressmaking fabric is sold by the metre. The bridesmaid's dress pattern requires 5 metres of pale pink voile. 5 metres of voile costs £57.50. The bride's dress requires 9 metres of the same fabric but in white.

What is the cost of 9 metres of white voile material?

A £103.50
B £107.50
C £111.50
D £115.00

10 The bride-to-be wants to lose 1.5 kilograms in the next three months. Her present weight is shown on the scale.

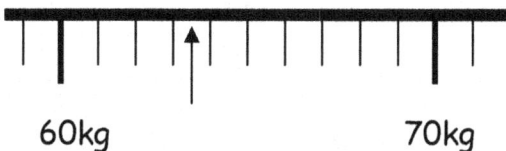

60kg   70kg

What is her target weight?

A 64.5kg
B 63.5kg
C 62.5kg
D 62.0kg

Questions 11 to 14 are about a travel agency.

A travel agent at HighLife Travel analyses sales of the different types of holiday packages bought during the Spring/Summer season.

Holiday Type	April	May	June	July	August	September	Total
Seaside Resorts	89	92	97	114	133	75	600
City Tours	57	58	47	69	40	29	300
Day Trips	72	95	63	75	57	46	408
Lakeland Weekends	39	53	65	88	107	98	450
Total:	257	298	272	346	337	248	1758

11 What is the mean number of **Day Trips** over the six-month period?

A 50
B 60
C 68
D 78

12 What is the range of the numbers of the different types of holiday in June?

A   32
B   34
C   45
D   50

13 The table shows the number of holidaymakers who have travelled with HighLife over a 4-year period.

HighLife	Number of Holidaymakers	
Year	Adults	Children
2004	225	150
2005	280	210
2006	290	220
2007	350	275

In 2004 what was the ratio of Adults to Children?

A   2 : 3
B   3 : 2
C   1 : 4
D   4 : 1

14 In 2007 what percentage of HighLife holidaymakers were Adults?

A   56%
B   67%
C   78%
D   89%

Questions 15 to 20 are about Valley Farm, a company that manufactures healthy snack bars.

15 A pack of the product Oatie Crunch snack bars gives nutritional information on the label:

Typical Values	Per 100g	Per 42g bar
Energy (Calories)	427kcal	179kcal
Protein	8.0g	3.4g
Carbohydrate	56.0g	23.5g
Polyunsaturated Fat	16.0g	6.7g
Fibre	6.0g	2.5g
Sodium	0.29g	0.12g

How much carbohydrate is there in Oatie Crunch compared to fat?

A   8 times as much
B   5 times as much
C   3.5 times as much
D   2.5 times as much

16 Oatie Crunch is baked on metal trays and later cut into bars by machine. The diagram shows a slab of Oatie Crunch before cutting.

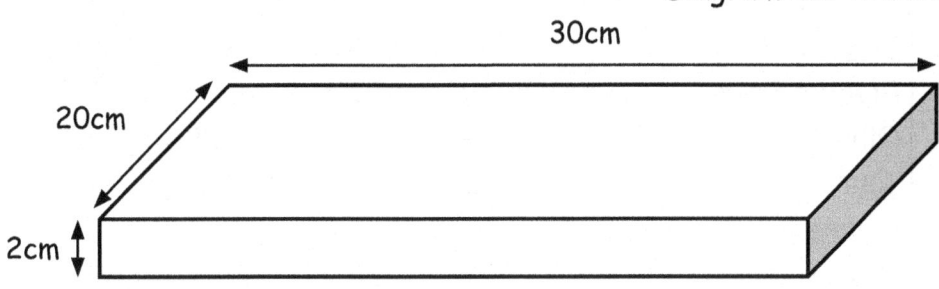

Diagram not to scale

The formula to calculate volume is length x width x height

What is the volume, in cubic centimetres, of Oatie Crunch in the slab before cutting?

A   100cm³
B   600cm³
C   800cm³
D   1200cm³

17 Individual Oatie Crunch bars measure 2cm x 5cm x 3cm. How many bars can be cut from one slab of Oatie Crunch?

A   50
B   40
C   30
D   20

18 Oatie Crunch bars are baked the oven at 350° Fahrenheit.

To convert from degrees Fahrenheit to degrees Celsius use the formula:

$$°C = \frac{(°F - 32) \times 5}{9}$$

What is the oven temperature, rounded to the nearest degree, in degrees Celsius?

A   157°C
B   176°C
C   177°C
D   187°C

19 A summary of payments and income for Valley Farm's trading over three months is shown in the table:

Valley Farm Quarterly Accounts Summary (£000s)				
Income	April	May	June	Totals
Product Sales	98.5	102.8	112.9	314.2
Other	4.7	6.3	9.6	20.6
Totals	103.2	109.1	122.5	334.8
Payments				
Raw Materials	20.6	22.9	23.9	67.4
Business Overheads	7.5	7.5	8.2	23.2
Wages	5.7	5.8	5.9	17.4
Totals	33.8	36.2	38.0	108.0

To verify the accuracy of the Total Income over the three months which two of the following check methods should be used?

Check method 1: $33.8 + 36.2 + 38.0 = 108.0$
Check method 2: $334.8 - 20.6 = 314.2$
Check method 3: $334.8 - 103.2 - 109.1 = 122.5$
Check method 4: $98.5 + 102.8 + 112.9 = 314.2$

A   Check methods 2 and 3
B   Check methods 3 and 4
C   Check methods 4 and 1
D   Check methods 1 and 2

20 What is the mean of the Total **Payments** for April, May and June?

A   £33 800
B   £36 000
C   £67 400
D   £108 000

Please go on to the next page

Questions 21 to 24 are about the number of computers and data storage devices in homes in a neighbourhood.

A consumer magazine carried out a survey to find out the number of computers that were in use in family homes on a housing estate. The results are shown in the table:

Consumer Magazine Survey		Number of children in the family				
		0	1	2	3	4
Total number of computers in the home (PCs and laptops)	0	2	1	0	0	1
	1	7	8	13	10	7
	2	5	7	11	4	7
	3	2	2	6	3	2

21 In how many of the families surveyed are there at least 2 children?

A 16
B 18
C 41
D 64

22 In how many homes are there 2 or more children **and** 2 or more computers?

A 33
B 31
C 18
D 16

23 An on-line computer company advertises a laptop computer that normally costs £499. In a special offer, the computer is reduced in price by 20%. What is the special offer price of the laptop?

A £99.80
B £199.20
C £299.80
D £399.20

24 Memory sticks for data storage normally sell for £1.19 each. In a 'buy one get one free' offer how many memory sticks does a customer get for a spending limit of £5.00?

A 4
B 5
C 8
D 9

Please go on to the next page

Questions 25 to 27 are about temperatures in European cities.

The table shows the temperature in various cities on one day in February.

City	Maximum Daytime Temperature (°C)
Bonn	-4
Brussels	4
Dublin	3
Edinburgh	1
Gdansk	0
Geneva	4
Leningrad	-11
Milan	5
Nicosia	17
Swansea	-2

The following day the Maximum Daytime Temperature in Leningrad increases by 4°C and the temperature in Bonn decreases by 2°C.

25 On that following day, by how many degrees warmer is Bonn than Leningrad?

   A   1°C
   B   5°C
   C   9°C
   D   13°C

The graph shows temperature variation in Dublin over one day in February:

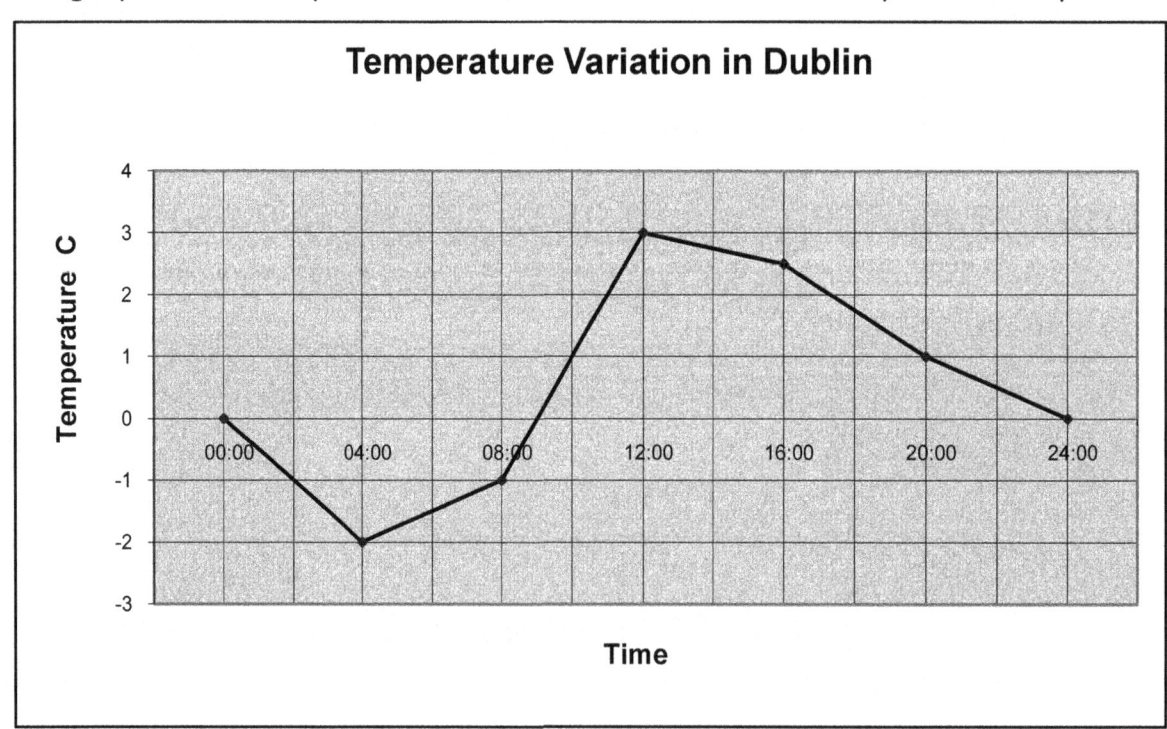

26 What is the closest approximation of the temperature in Dublin at 4pm?

    A   -2°C
    B   0°C
    C   1.5°C
    D   2.5°C

On one day in Swansea a thermometer shows a temperature of 30 degrees Fahrenheit.

> To convert degrees Fahrenheit to degrees Centigrade use the formula:
> $$°C = \frac{(°F-32) \times 5}{9}$$

27 What is the temperature, to the nearest degree in Centigrade, in Swansea?

    A   1°C
    B   -1°C
    C   11°C
    D   -11°C

The diagram shows a picture frame drawn to the scale of 1:20.

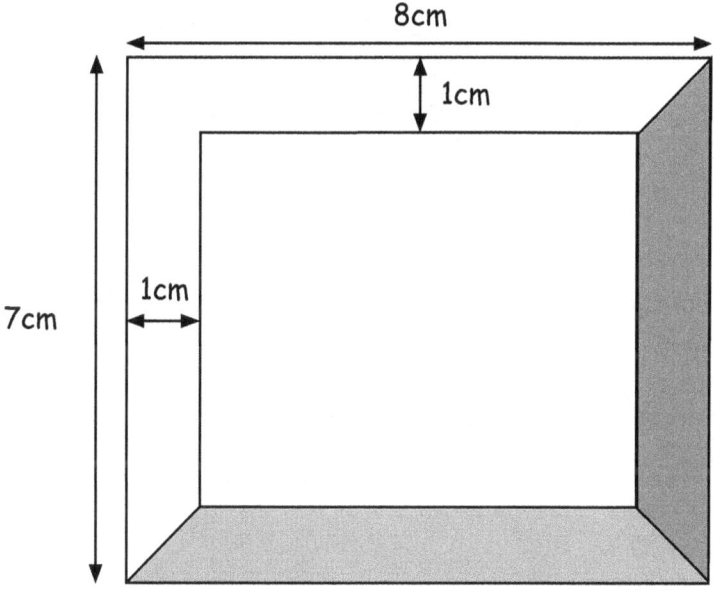

28 What are the approximate dimensions of the full-sized picture space inside the frame?

    A   1400mm x 1600mm
    B   1200mm x 1600mm
    C   1200mm x 1400mm
    D   1000mm x 1200mm

Questions 29 and 30 are about foreign currency exchange.

29 Janet's bank charges no commission on foreign currency exchange when ordered in advance. She wishes to change £750 to Euros.

The exchange rate is currently gives €1.6 for every £1.

To the nearest Euro, how many Euros does Janet receive?

A   €469
B   €1200
C   €1500
D   €1620

30 The pie chart shows the three most popular currencies exchanged for pounds at Janet's bank this month: American Dollars, Euros and Japanese Yen.

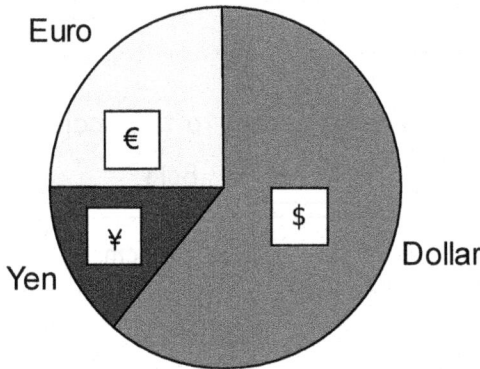

The total number of pounds exchanged for foreign currency this month is £2.8 million.

How many pounds were exchanged for Euros?

A   £1,700,000
B   £1,400,000
C   £700,000
D   £400,000

Please go on to the next page

Questions 31 to 34 are about updating the gardens of a bungalow.

31 A garden designer draws a diagram of a lawn section with a semi-circular edge and a herbaceous border for the rear garden:

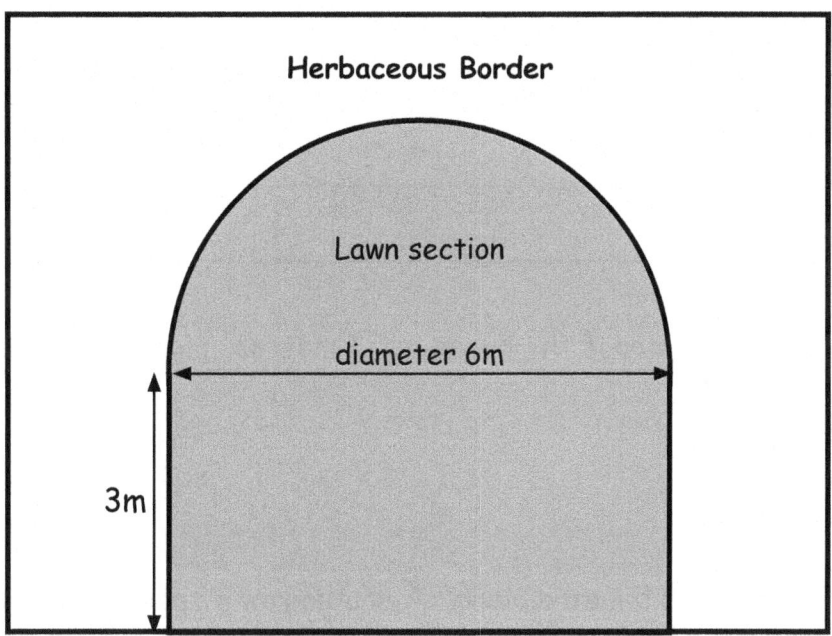

Diagram not to scale

A formula to work out the approximate area of a full circle is:

Area = 3 × (radius)^2

Which calculation works out the total area, in square metres, of the lawn section?

A $(3 \times 6) + \dfrac{1}{2} \times (3 \times 3^2)$

B $(3 \times 6) + 3 \times 6^2$

C $(3 \times 6) + 3 \times 3^2$

D $(3 \times 6) + \dfrac{1}{2} \times 9^2$

32 Instructions on the label of a 10-kilogram bag of fertiliser recommends coverage of 250 grams per square metre. The herbaceous border is 1.5 metres wide.

Which calculation works out the length of border that can be covered by one bag of fertiliser?

A $(10000 \div 1.5) \times 250$

B $(10000 \div 250) \times 1.5$

C $(10000 \div 250) \div 1.5$

D $10000 - (1.5 + 250)$

A large, mature tree in the front garden is shown in the diagram in relation to the height of the bungalow.

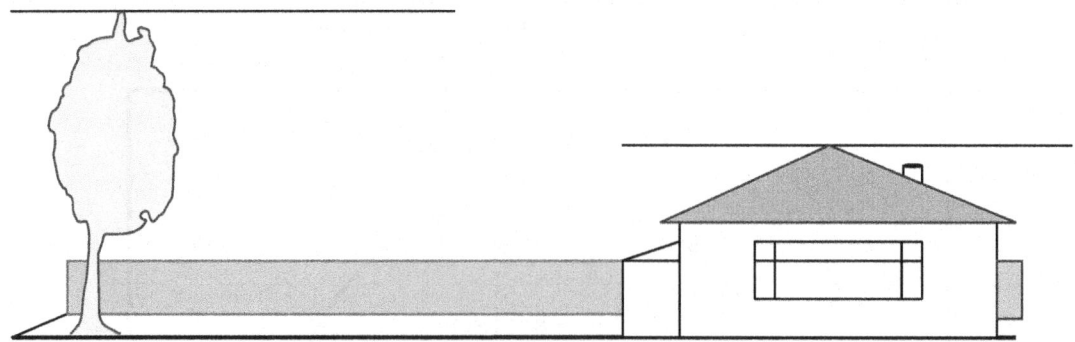

33 To the apex of the roof the house is 5.1 metres. How tall is the tree?

  A 10.2 metres
  B 8.6 metres
  C 8.0 metres
  D 7.6 metres

34 The driveway is 18.5 metres long. A double row of paving slabs is to be laid leading up to the garage. Each paving stone measures 0.75 metres by 0.75 metres. How many **whole** paving slabs are needed to complete the job?

  A 25
  B 36
  C 48
  D 50

Questions 35 and 36 are about a zoo.

35 In the reptile house a zoo keeper records the lengths, in centimetres, of 20 baby alligators:

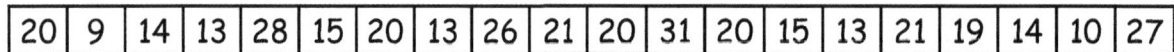

| 20 | 9 | 14 | 13 | 28 | 15 | 20 | 13 | 26 | 21 | 20 | 31 | 20 | 15 | 13 | 21 | 19 | 14 | 10 | 27 |

What is the median length of the baby alligators?

  A 17.5cm
  B 18.0cm
  C 19.5cm
  D 20.0cm

Please go on to the next page

36 The chart shows the number of visitors on each day of one week:

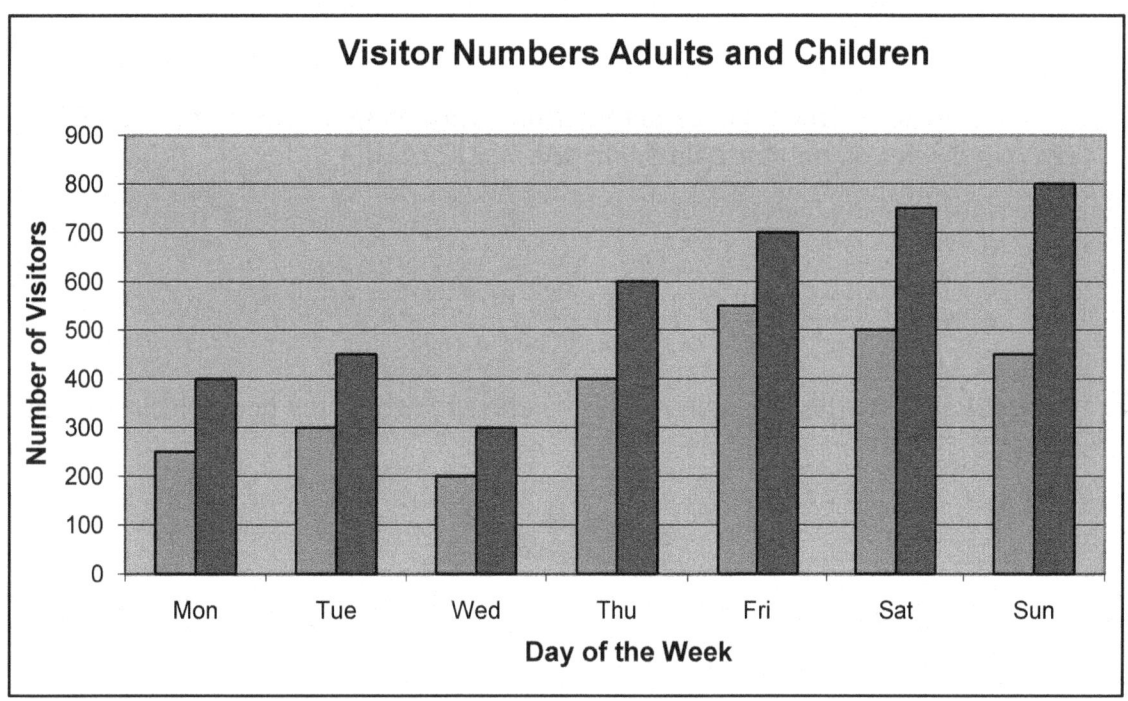

What is wrong with this chart?

- **A** the horizontal x-axis is not labelled
- **B** the bars are not plotted correctly
- **C** the key is missing
- **D** the vertical y-axis is missing

Questions 37 to 40 are about a sports club.

37 Pelton Tri-racquet club has several courts for playing tennis, squash and badminton and a membership of 240 players. 20% of members are in the Youth group for players under 18 years of age.

How many club members are in the Youth group?

- **A** 24
- **B** 48
- **C** 60
- **D** 72

38 During the summer holidays the Youth group books the tennis courts for two-fifths of the available playing time. What percentage of the available playing time is booked by the Youth group?

- **A** 20%
- **B** 25%
- **C** 30%
- **D** 40%

39 The concentrated form of a high-energy isotonic drink, popular with tennis players, is sold in 75cl bottles. The concentrate is diluted with water in the ratio of 1 : 7.

When diluted in the correct proportions, how many litres of dilute high-energy drink can be made up from one 75cl bottle?

A   7 litres
B   6 litres
C   5 litres
D   4 litres

40 At weekends in the summer season fresh strawberries are sold in the club house.

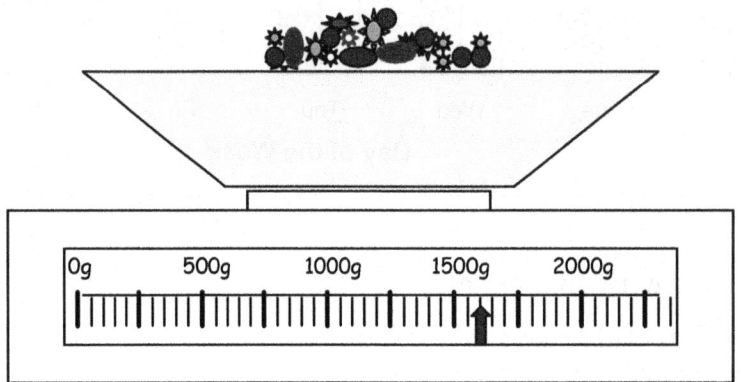

What is the weight of the strawberries on the scale?

A   1kg 625g
B   1kg 600g
C   1kg 550g
D   1kg 525g

End of Paper

Practice Multiple-choice Paper
Suitable for:

Key Skills Level 2 Application of Number
Level 2 Adult Numeracy

Paper Eight

YOU NEED

- This test paper.
- A pen.
- A pencil and eraser.
- An Answer Sheet.
- A ruler marked in centimetres and millimetres

You may NOT use a calculator.
You may use a bilingual dictionary.
There are 40 questions on this paper. Try to answer ALL the questions.
When you have completed the questions you must check your answers, then check them again.

YOU HAVE QUARTER OF AN HOUR TO READ THE PAPER
AND ONE HOUR TO COMPLETE THE 40 QUESTIONS

INSTRUCTIONS

- Make sure you write your name and today's date on the Answer Sheet. Use a pen to do this.
- Use a pencil to mark your answers so if you change your mind you can erase your choice and select another.
- Make sure that for each question you have only selected one answer. If you select more than one, the answer will not be marked.
- Read each question carefully before you select an answer.

Note for learners and tutors: This is a practice test that has been designed to closely resemble the questions and question styles of a "live" paper.

Questions 1 to 5 are about an independent bakery business.

1   A baker records the number of loaves he sells, Monday to Saturday, during February. To make a profit he must to sell **more than** 640 loaves per week.

Loaves Sold				
February	Week 1	Week 2	Week 3	Week 4
Monday	667	642	621	627
Tuesday	647	640	679	690
Wednesday	643	693	611	642
Thursday	640	613	621	628
Friday	645	623	679	645
Saturday	650	694	708	700
Total	3892	3905	3919	3931

On how many days in February does the baker make a profit from loaves?

A   17
B   15
C   9
D   4

2   What is the range of the number of loaves sold in February?

A   97
B   95
C   89
D   87

3   The baker supplies local shops with bread rolls at the prices shown in the table.

Type of Roll	Price
White Floury Bap	11p
White Crusty Roll	12p
Wholemeal Bun	13p
Granary Bun	13p
Torpedo	15p

The shops add 35% to the cost of each roll to sell to customers. Which calculation works out the cost of one Granary Bun to a customer?

A   13 + 35/100
B   13 × 35/100
C   13 – (35/100 × 13)
D   13 + (35/100 × 13)

4   To help him with his business accounting, the baker purchases a computer. The cash price of the computer system is £720. Using his credit card he pays a deposit of £200 and agrees to pay £49.99 in 12 monthly instalments.

To the nearest pound, how much above the cash price does he pay by using his credit card?

A   £104
B   £80
C   £79
D   £69

5   The baker compares his product sales for 2006 and 2007 in a bar chart:

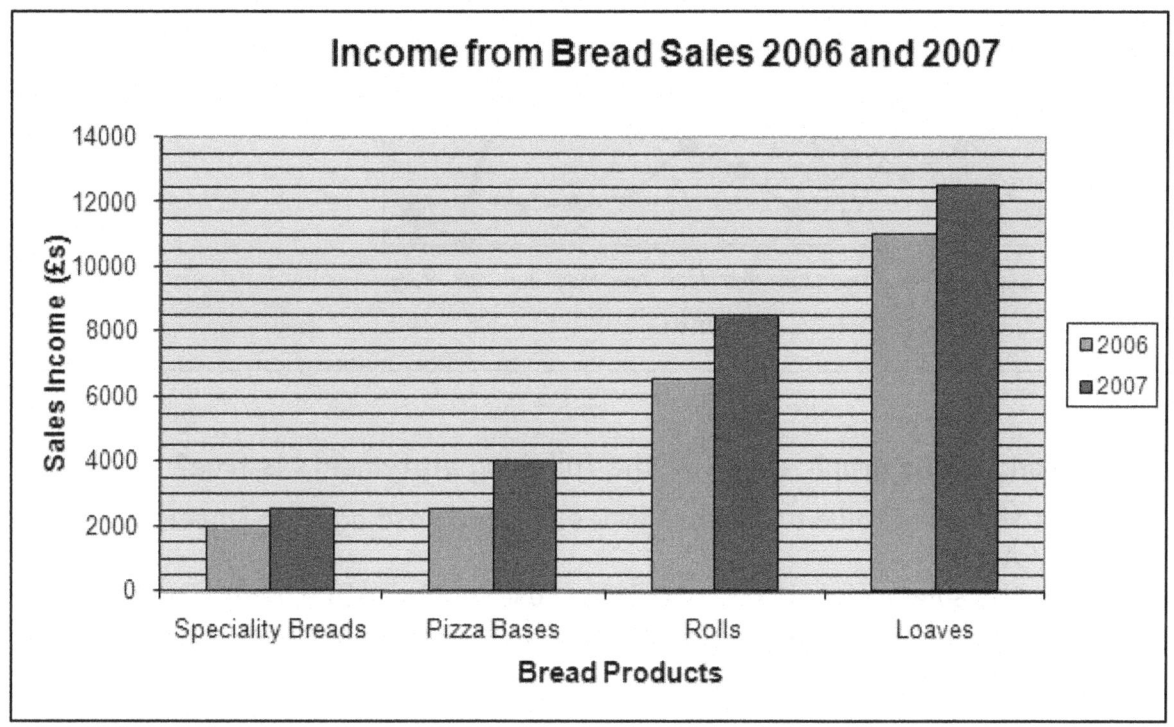

Comparing the sales figures for 2006 and 2007, which of the following statements is correct?

A   the income from Loaves increased by £3000
B   the income from Speciality Breads decreased by £500
C   the income from Pizza Bases increased by £1500
D   the income from Rolls increased by £4000

Please go on to the next page

Questions 6 and 7 are about the bookshop on a university campus.

6   The bookshop manager draws a graph to show the value of sales from computer consumables, stationery items, photocopying and university logo merchandise.

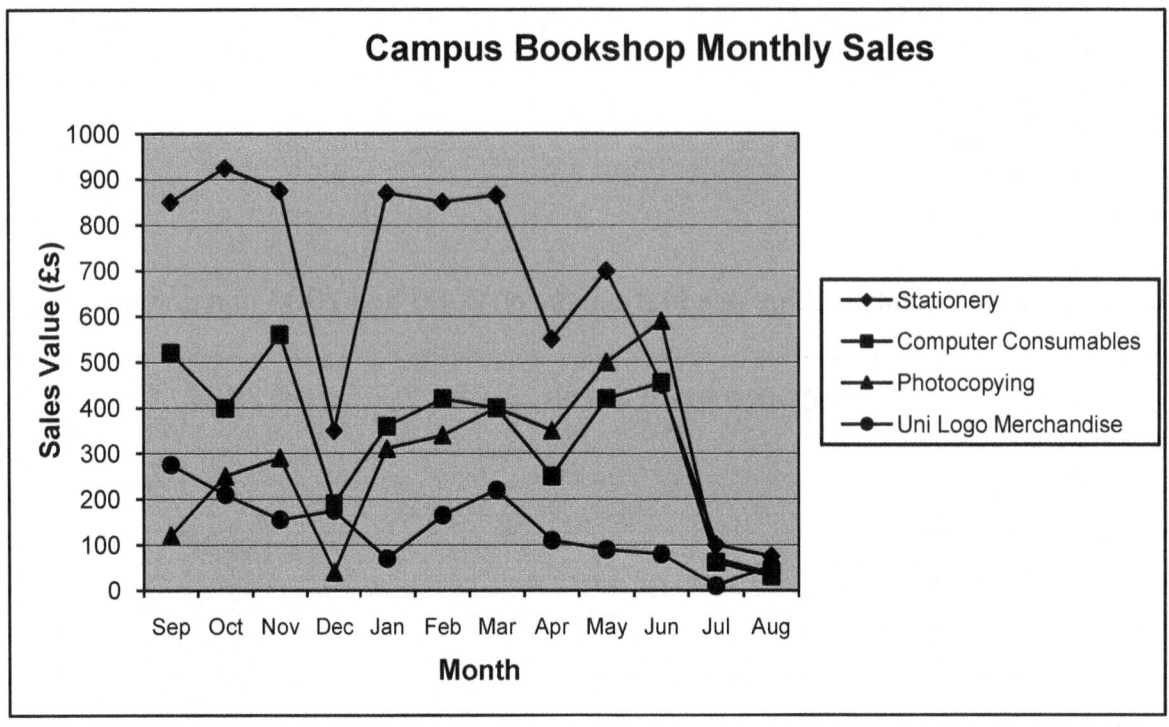

      Looking at the graph, which of the following statements is true?

      A   the sales value of all categories fell and rose again in December and April
      B   the sales value of all categories rose in November, March and June
      C   the sales value of Photocopying is lower than that for Stationery every month
      D   the sales value of stationery is higher than University Logo Merchandise every month

7   The bookshop manager wants to work out the average monthly running costs of the business. She does not want the average to be distorted by unusually high or low running costs.

      Which is the best measure of the average she can use?

      A   the range
      B   the median
      C   the mean
      D   the mode

Questions 8 to 12 are about Zippy's Fast Food Franchise.

8 Students Bradley (19 years old), and his sister, Rebecca (17 years old) work part time hours at Zippy's during term time. The table shows minimum wage rates for employee age ranges.

Minimum Wage Rates (per hour)	
16 – 17 years	£3.40
18 – 21 years	£4.60
22 years and over	£5.52

One week Bradley and Rebecca work the same hours:

Evening Shift	Hours worked
Thursday	3
Friday	3
Saturday	4
Sunday	2.5

How much more is Bradley paid than Rebecca for working the same number of hours?

A £26.75
B £15.00
C £12.50
D £11.00

9 Three and a half litres of orange juice are sold during one morning shift. Each glass contains 175ml of juice and costs 65p. Which calculation gives the total income in pounds from the sale of orange juice?

A 3.5 x 175 x 0.65
B 3500 x 175 x 0.65
C 3500 ÷ 175 x 0.65
D 3500 ÷ 1.75 x 65

10 An assistant supervisor at Zippy's currently earns £18,750 per annum. She receives a 3% salary increase on her successful completion of a qualification.

Which calculation finds the newly increased monthly salary amount?

A 18750/12 + (18750/12 x 3/100)
B 18750/12 + (18750/12 x 100/3)
C 18750/12 + (18750 x 3/100)
D 18750 + (18750/12 x 3/100)

A supervisor creates a chart showing a comparison of the sales of Burgers and French Fries.

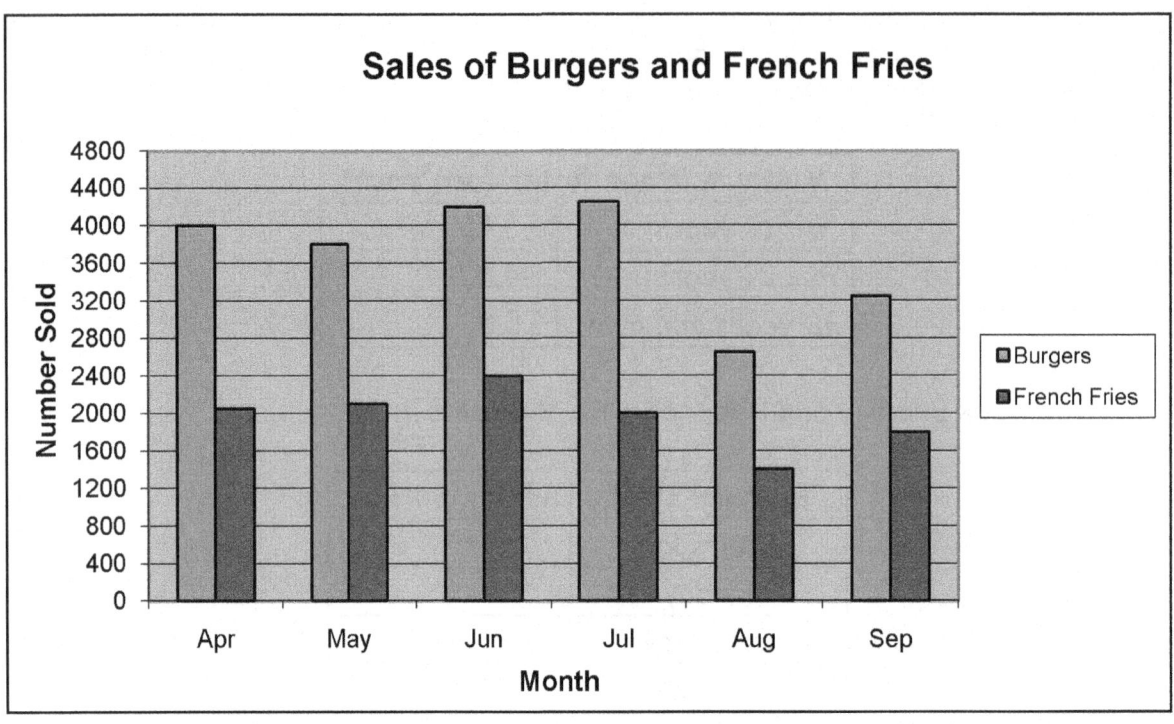

11   What is the difference between the number of French Fries sold in June and in August?

A   1300
B   1200
C   1000
D   800

The manager checks a credit account statement for one month's trading:

Date	Description	Amount
	Previous Balance Carried Forward	1332.20 -
13.08.08	Allied Butchers, Penton-lee	584.75 -
15.08.08	Stubbs Potato Merchants	321.50 -
22.08.08	Greenway Vegetable Supplies	289.90 -
01.09.08	Villa Soda Co.	178.25 -
04.09.08	Payment received with thanks	2000.00 +
11.09.08	Interest charge	17.75 -
	New Balance Owing	

12   What is the amount of the New Balance Owing?

A   £52.20
B   £69.95
C   £696.60
D   £724.35

Questions 13 and 14 are about 'Pet Meals' animal feed delivery service.

13   Pet Meals deliver animal feeds to a local animal shelter. Prices, in £s, of the most popular animal feeds are shown in the table:

Animal Feed (Dry)	Medium	Large	Economy	Jumbo
Horse	13.50	26.50	75.00	100.00
Dog	8.00	15.50	19.00	27.50
Cat	7.50	14.50	18.00	25.00
Rabbit	5.25	10.50	15.00	20.00

The shelter manager takes delivery of an order and the total bill is £140.25. The order was for two Jumbo bags of cat food and two Jumbo bags of dog food and also Rabbit food.

What quantity and what size bags of rabbit food were included in the order?

A   4 medium
B   3 large
C   1 jumbo and 1 medium
D   2 economy and 1 medium

14   The fuel tank in the Pet Meals delivery truck contains 12 gallons when full. The emergency reserve can holds one-tenth of the fuel tank's volume.

One gallon = 4.55 litres

Which calculation works out the capacity of the reserve can in litres?

A $(12 \times 1/10) \div 4.55$
B $(12 \times 1/10) \times 4.55$
C $(12 \times 9/10) \div 4.55$
D $(12 \times 9/10) \times 4.55$

Questions 15 and 16 are about a party of tourists visiting the Grand Canyon in the USA.

A group of 12 adult tourists wish to take the helicopter tour of the Grand Canyon during a holiday in Arizona. The cost of helicopter trips are shown in the table:

Grand Canyon Tour Trip Tariff		
Per Person	One Hour	Two Hour
Adult	$69	$99
Child	$49	$79
Group discount of 10% for parties of 10 persons or more		

15 To the nearest dollar, what is the total cost of a one-hour trip for 12 adults?

A $1188
B $1069
C $828
D $745

> **The exchange rate for British pounds to American dollars is:**
> **$1.9 to £1**

16 How much does it cost, in British pounds, for a two-hour trip around the Grand Canyon for two adults?

A £124.20
B £104.21
C £83.16
D £76.41

Questions 17 to 19 are about Newton College sport activities.

17 The Newton College basketball team is in a regional league. The table shows the top five teams' performance after six matches so far this season.

Northern Basketball League Table								
	Matches this season				Goals scored			
	Played	Won	Drawn	Lost	For	Against	Goal difference	Points
Huyton	6	4	1	1	157	77	80	
Newton	6	3	1	2	145	134	11	
Belham	6	2	1	3	103	145	-42	
Toreham	6	2	0	4	99	137	-38	
Merton	6	2	0	4	97	140	-43	

Teams gain two league points for each match they win and one league point for a draw. How many league points have Newton gained so far?

A 5
B 7
C 9
D 10

Please go on to the next page

18   Twenty cyclists take part in time trials to represent Newton College at regional level. The race steward records each cyclist's speed at the start of the last lap of the sport stadium:

16.3	16.6	14.4	17.6	15.4
13.4	15.2	15.2	14.9	15.1
15.7	17.8	17.4	19.3	15.4
14.8	19.8	14.3	18.3	15.2

The steward rounds up the recorded speeds and starts to put them into a table:

Newton College Cycling Race Time Trials								
Speed (km per hour)	13	14	15	16	17	18	19	20
Number of cyclists	1	2	8	2				

Looking at the first table, how many of the cyclists will the steward record as having achieved a rounded speed of 17 kilometres per hour?

A   4
B   3
C   2
D   1

19   A coach is hired to transport the basketball team and support personnel to a tournament. The coach carries 33 passengers and costs £234.23 . To estimate the approximate cost per person the team manager rounds up each of the figures to the **nearest 10**.

Which calculation does he use?

A   230 ÷ 30 = £8.00
B   235 ÷ 33 = £7.00
C   240 ÷ 40 = £6.00
D   230 ÷ 40 = £5.75

Questions 20 to 25 are about the Eco-Village tourist attraction.

20   The marketing manager at Eco-Village keeps records of visitor numbers per day. He wants to display this information for one week. Which display method is the most appropriate?

A   a scatter graph
B   a scale drawing
C   a pie chart
D   a bar chart

21 A guesthouse near the Eco-Village site is fully booked for the summer season.

Eden Vale Guesthouse, St Columbus Road, Estham, ES2 8FG		
Number of rooms	Guestroom description	Cost per person per night
2	Double bedroom with en-suite bathroom	£60
3	Twin bedded room with en-suite bathroom	£50
2	Single bedroom with shower facility	£45
Room rates inclusive of full English or continental breakfast		

What is the total weekly income when all the rooms are booked Monday to Sunday at the Eden Vale Guesthouse?

A   £4410
B   £2100
C   £1680
D   £630

22 Valley View offers Bed and Breakfast accommodation to Eco-Village visitors. The table shows the landlady's record of rooms booked during one week in October. During the off-peak season the landlady only hires extra staff on days when there are **eight or more** guests booked in.

Valley View B&B Booking Dates in October							
Room No	4th	5th	6th	7th	8th	9th	10th
1	2	0	2	2	2	2	0
2	0	2	1	1	2	2	2
3	2	2	1	1	2	2	1
4	2	1	2	0	2	0	2
5	1	0	1	1	1	1	1
6	0	1	1	0	1	1	0
7	1	0	0	0	1	1	0

On which dates are there 8 or more guests?

A   4th, 6th and 8th
B   4th, 8th and 9th
C   6th, 8th, 9th and 10th
D   4th, 6th, 8th and 9th

23 The mean monthly rainfall for the area where the Eco-Village is located is 70 millimetres. The table records monthly rainfall for 2006 and 2007.

Rainfall	Jan	Feb	Mar	Apr	May	Jun	Jul	Aug	Sep	Oct	Nov	Dec
2006	80	43	135	67	26	20	41	45	47	105	35	49
2007	79	64	23	139	70	25	61	108	78	120	99	121

In how many months in the two years was the rainfall **more than** 10mm above mean monthly rainfall?

A   8
B   7
C   5
D   4

24 A stand in the Education Centre at the Eco-Village has a display of the amount of raw materials imported into the country which could be saved if we increase recycling.

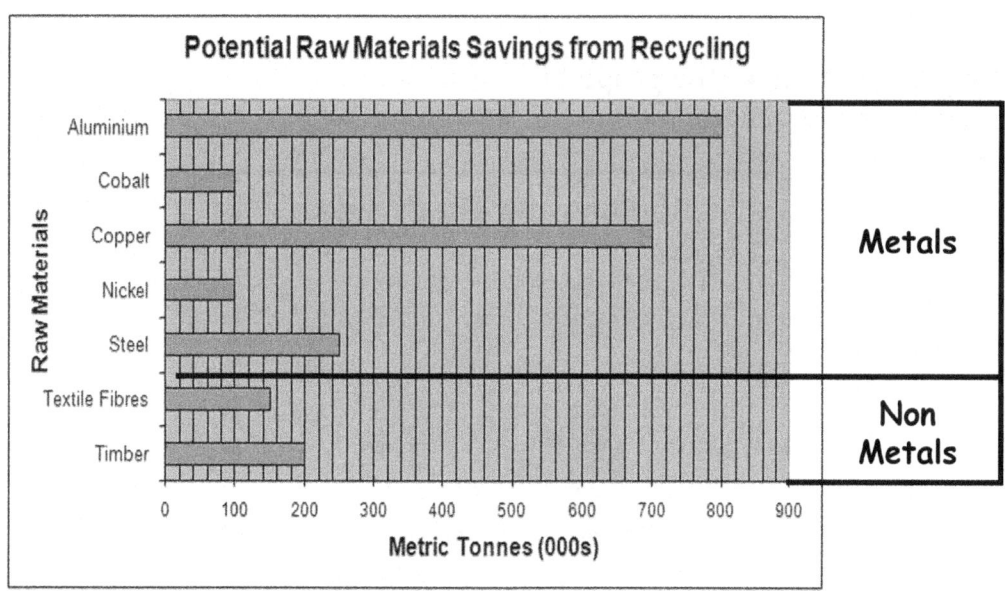

What is the total, in metric tonnes, of non-metal imports?

A 200 metric tonnes
B 800 metric tonnes
C 350,000 metric tonnes
D 800,000 metric tonnes

25 The Eco-Village Education Centre promotes wind power generation. A display board shows wind power generation figures for a number of European countries.

Wind Power Generation - Northern Europe		
Country	Population (millions)	Wind Power (megawatts)
Germany	82.3	2231
Denmark	5.3	2310
France	61.7	156
Netherlands	16.1	635
Norway	5.4	127
Sweden	9.2	340
United Kingdom	62.4	583

Referring to the table, which one of these statements is true?

A The UK generates more wind power per million population than Germany.
B The country with the largest population generates the least wind power.
C Denmark generates less wind power per million population than Norway.
D The country with the smallest population generates the most wind power.

Questions 26 and 27 are about a shop selling curtains and blinds.

26 B. W. Jefferson & Co supplied curtains and blinds to 6,000 customers last year.

The pie chart shows the proportion of customers who bought various types of curtains and blinds.

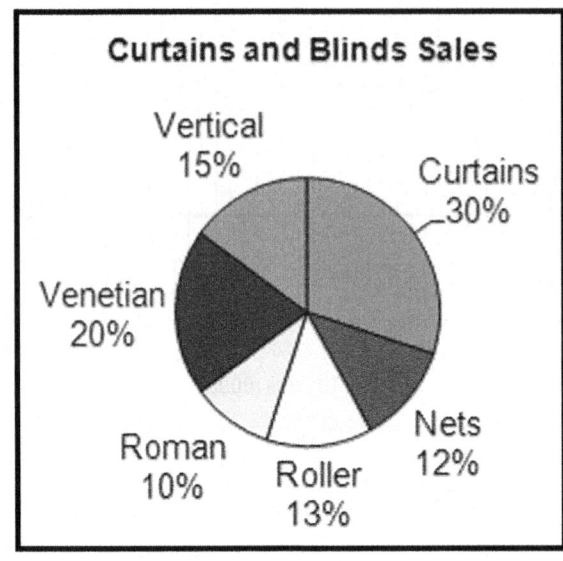

How many customers bought Vertical Blinds?
A   600
B   720
C   900
D   1200

27 The table below shows the prices of curtains, tiebacks and matching cushion covers:

B. W. Jefferson & Co. Curtain and Blinds Suppliers	
Design: Swirl Devore, Cream and White, 200cm wide	
Curtain Length	Price
120 cm	£28.75
150 cm	£39.75
180 cm	£59.45
Matching Items	
1 pair Tie-backs	£4.95
Cushion Covers: each	£5.95

A customer orders a pair of curtains, 180 cm in length plus a pair of tiebacks and two matching cushion covers. Which calculation gives the nearest estimate, in pounds, of the total cost of the order?

A   60 + 6 + (2 x 5)
B   60 + 5 + (2 x 6)
C   59 + 6 + (2 x 5)
D   59 + 5 + (2 x 6)

Questions 28 and 29 are about Esham Vale Church Hall.

28 Esham Vale Church Hall hires out the main hall to community groups for events. The diagram shows a plan of the main hall with measurements.

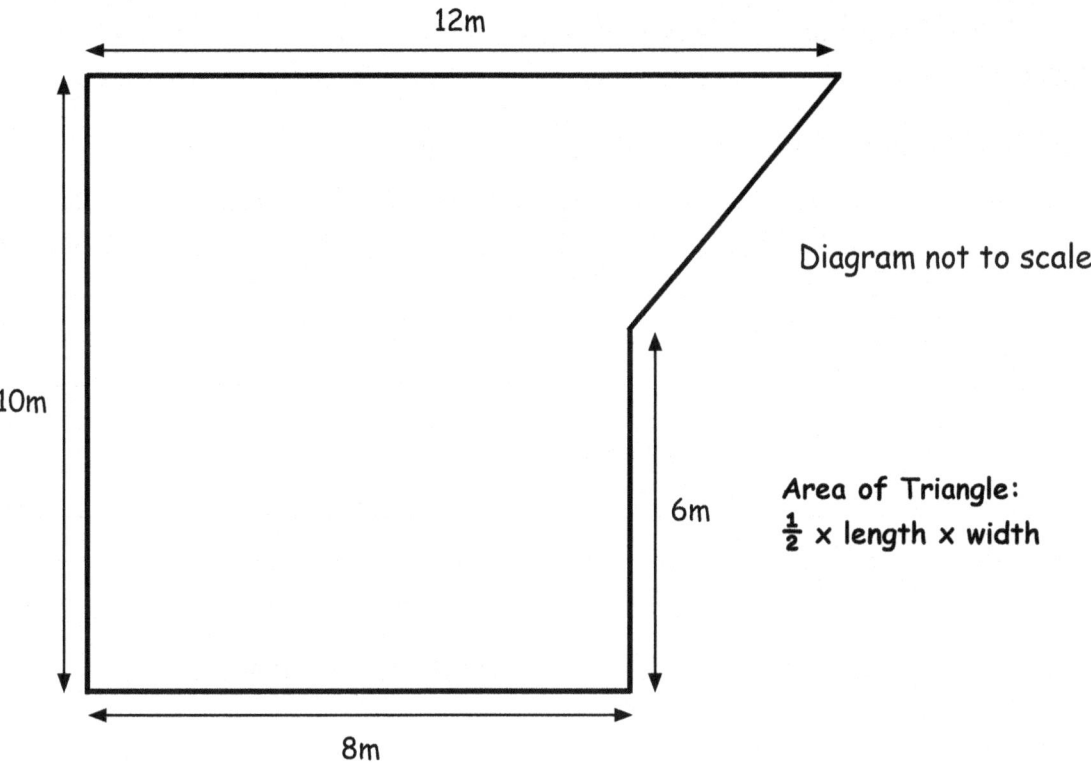

Diagram not to scale

Area of Triangle:
$\frac{1}{2}$ x length x width

What is the total floor area of the main hall?

A 80m²
B 88m²
C 96m²
D 120m²

29 The treasurer of Esham Vale Church committee works out the total cost of 150 gallons of kerosene heating oil for the Church central heating system.

The cost per litre of oil is 50.95p (i.e. fifty point nine five pence per litre)

1 gallon = 4.55 litres

Which calculation works out the total cost of oil in pounds?

A 150 x 4.55 x 50.95/100
B 150/4.55 x 50.95/100
C 150 x 4.55 x 50.95 x 100
D 150/4.55 x 50.95 x 100

Questions 30 to 32 are about a computer fair to be held at Esham Vale Church Hall.

30　The Big Byte computer magazine hosts a computer fair to be held in Esham Vale Church Hall. Big Byte wishes to hire the hall for **six hours** on a **Saturday** including the setting up of tables and chairs, but no kiosk required.

Charges for hiring the hall are shown in the table:

Esham Vale Church Hall Private Function Rates	
Hire Periods	Hire Rate per Hour
Monday to Wednesday	£25
Thursday and Friday	£35
Saturday	£55
Additional Charges	
Setting up tables and chairs	£30 per booking
Use of refreshments kiosk	£50 per booking
A deposit of 35% of the total cost is required in advance	

Which of the calculations works out the total deposit amount?

A　(30 + 55) × 6 × 100/35
B　(30 + 55) × 6 × 35/100
C　(55 × 6 + 30) × 100/35
D　(55 × 6 + 30) × 35/100

31　A stall holder at the fair buys in computer games CDs at £6.00 and sells them on for £16. Which calculation works out the percentage profit made on each CD?

A　16 ÷ 6 × 100 = 266.66%
B　(16 – 6) ÷ 6 × 100 = 166.66%
C　(16 – 6) ÷ 16 × 100 = 62.50%
D　16 ÷ 100 × 6 = 0.96%

32　On the day of the computer fair a total of 312 people visited, of this number, 96 are children. What was the ratio of adults to children?

A　4 : 1
B　1 : 4
C　4 : 9
D　9 : 4

Questions 33 to 34 are about a sponsored charity fun-run.

33 Janet, a fun-runner, receives sponsorship pledges from friends and family to take part in a ten mile fun-run:

11 sponsors pledge 50p per mile
7 sponsors pledge 20p per mile
12 sponsors pledge 10p per mile

Which calculation can she use to work out the total amount, in pounds, that she can raise for charity assuming that she finishes the 10 mile fun-run?

A $(11 \times 0.50) + (7 \times 0.20) + (2 \times 0.10) \times 100$
B $(11 \times 0.50) + (7 \times 0.20) + (12 \times 0.10) \times 10$
C $(11 \times 0.50) \times (7 \times 0.20) \times (12 \times 0.10) \times 100$
D $(11 + 7 + 12) \times (0.5. + 0.20 + 0.10) \times 10$

34 The fun-run supervisor needs a rough estimate of the amount of money to be raised so he asks each fun-runner how much they have been pledged. Which method is the most effective to record this information quickly?

A plotting points on a line graph
B drawing bars on a bar chart
C making tallies on a tally chart
D drawing segments on a pie chart

Questions 35 and 36 are about changing trends in motor vehicle transport.

35 The table shows the percentage change in the numbers of sport utility vehicles sold by car dealerships in various parts of the UK in the past year. In how many of the UK regions did the sales of sport utility vehicles decrease by at least 3%?

UK Region	Percentage change in sales
Cleveland	+ 3.3
East Anglia	- 5.4
East Midlands	- 3.0
North East	+ 7.1
North West	+ 4.0
South East	- 7.2
South West	+ 0.5
Greater London	- 2.5
West Midlands	- 4.2

A 3
B 4
C 6
D 7

36   The graph shows an upward trend in the average number of vehicles using a toll road tunnel per week between 2002 and 2008.

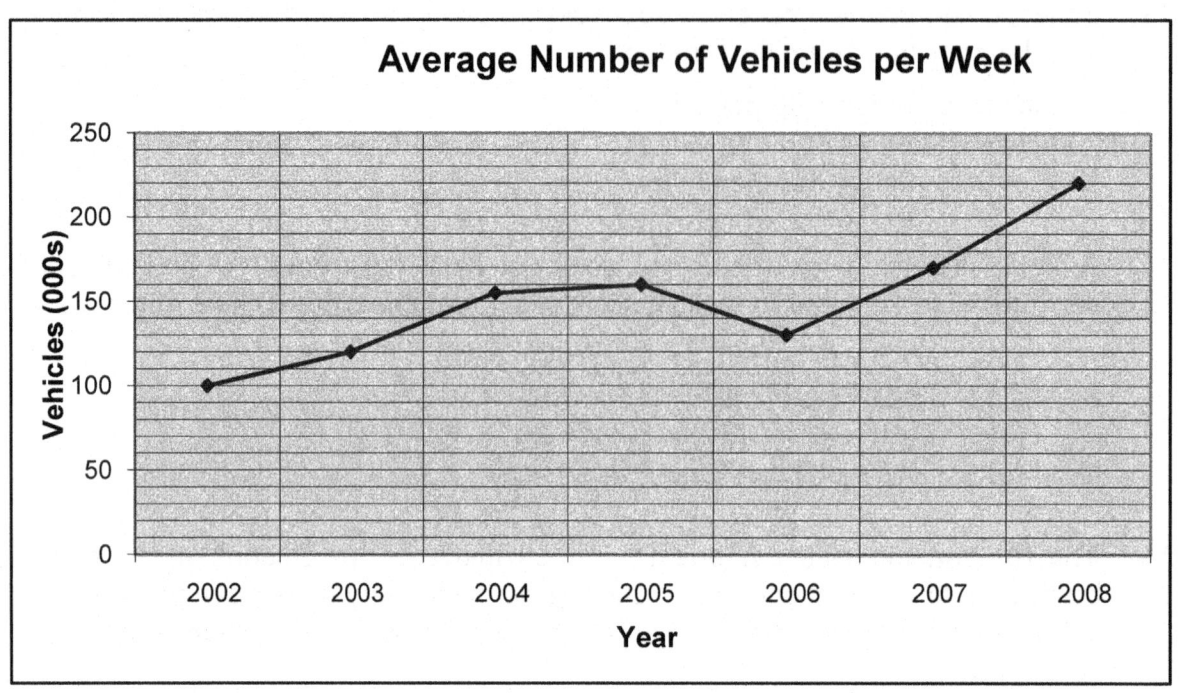

By what amount has the average weekly vehicle usage of the toll road tunnel increased between 2002 and 2008?

A   230,000
B   220,000
C   130,000
D   120,000

Questions 37 to 40 are about a college theatre group production.

37   Newton College Drama Studies department hires a rehearsal room at the Garland Theatre on Monday and Wednesday each week during term time for **30 weeks** of the year.

The table shows the theatre's hire costs per day:

Day	Charge per Day
Monday Tuesday	£70
Wednesday Thursday	£90
Friday	£120
Administration charge: £20 per Day	

What is the **total cost** to the Drama Studies department **per year**?

A   £6000
B   £5400
C   £1740
D   £540

38 The sixth form drama group are putting on a play at the Garland Theatre at the end of term. A week before the play is due to open only 5/8 of the tickets have been sold. What percentage of tickets remains **unsold**?

A   75.0%
B   62.5%
C   37.5%
D   30.0%

39 The performance space at the Garland Theatre is open plan and circular. The diameter of the performance area is 24m.

> Area of a circle is approximately 3 x r² **(r = radius)**

What is the area of the performance space?

A   72m²
B   432m²
C   864m²
D   1728m²

40 Choc-ices are on sale in the interval of the play.

Each choc-ice measures: 8cm x 5cm x 2.5cm. The container is 10cm deep but there must be 2cm clearance at the top.

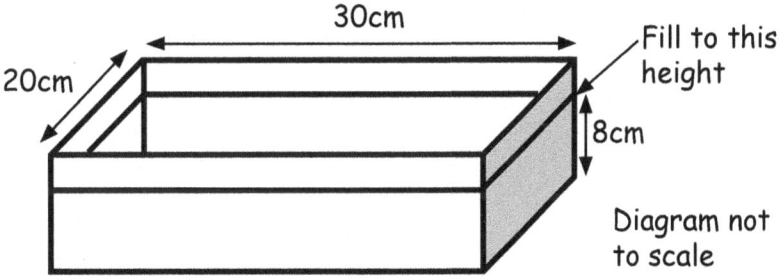

30cm

20cm

Fill to this height

8cm

Diagram not to scale

What is the maximum number of choc-ices that will fit into the container up to the fill height?

A   96
B   72
C   54
D   48

End of Paper

Practice Multiple-choice Paper
Suitable for:

Key Skills Level 2 Application of Number
Level 2 Adult Numeracy

Paper Nine

YOU NEED

- ■ This test paper.
- ■ A pen.
- ■ A pencil and eraser.
- ■ An Answer Sheet.
- ■ A ruler marked in centimetres and millimetres

You may NOT use a calculator.
You may use a bilingual dictionary.
There are 40 questions on this paper. Try to answer ALL the questions.
When you have completed the questions you must check your answers, then check them again.

YOU HAVE QUARTER OF AN HOUR TO READ THE PAPER
AND ONE HOUR TO COMPLETE THE 40 QUESTIONS

INSTRUCTIONS

- ■ Make sure you write your name and today's date on the Answer Sheet. Use a pen to do this.

- ■ Use a pencil to mark your answers so if you change your mind you can erase your choice and select another.

- ■ Make sure that for each question you have only selected one answer. If you select more than one, the answer will not be marked.

- ■ Read each question carefully before you select an answer.

Note for learners and tutors: This is a practice test that has been designed to closely resemble the questions and question styles of a "live" paper.

Questions 1 to 3 are about a Ten Pin Bowling Alley.

1   Flips Bowling Alley offers special ticket rates to members during summer holidays. Between 10am and 4pm, a child ticket costs £1.60. The adult ticket costs 80p **more** than a child ticket.

Compared to the cost of an adult ticket, what fraction is the cost of a child ticket?

A   1/3
B   2/3
C   3/5
D   4/5

2   In June, out of 1225 people who visited Flips Bowling Alley 700 were adults, the rest were children. What is the ratio of adults to children?

A   9 : 5
B   5 : 9
C   3 : 4
D   4 : 3

3   In July, visitor numbers to Flips Bowling Alley increased from the June figure of 1225 by 245.

What is the percentage increase in visitor numbers between June and July?

A   15%
B   20%
C   30%
D   50%

4   In one hour at the bowling alley, 21 pairs of ladies' bowling shoes were hired out. The size and frequency of hiring each shoe size are shown in the table.

Shoe size	3	3½	4	4½	5	5½	6	6½	7	7½	8	8½
Frequency	3	2	4	2	2	1	0	2	1	1	2	1

What is the median number of the sizes of ladies bowling shoes hired out?

A $6\frac{1}{2}$

B   5

C $4\frac{1}{2}$

D $3\frac{1}{2}$

Please go on to the next page

Questions 5 to 9 are about a cycling holiday in The Netherlands.

5   John and his son, James, plan a cycling tour in The Netherlands and wish to take their own bicycles with them on the ferry to Rotterdam. The ferry times and ticket prices are shown in the table.

Rotterdam Ferry	Pricing Periods:				(Prices per passenger)	
Outward Journey Dates:	31 Mar – 27 Apr 16 Jul – 19 Aug		28 Apr – 15 Jul 20 Aug – 28 Sep		29 Sep – 30 Apr	
Ticket Type	Single	Return	Single	Return	Single	Return
Short Break 5 days or less	£91	£172	£68	£126	£44	£98
Super Saver 6 to 14 days	£133	£256	£92	£174	£66	£122
Standard Fare 15 days or more	£169	£328	£116	£222	£76	£142
Bicycle Stowage per bicycle	£25	£40	£22	£34	£20	£30

John and James plan to depart on July 17th and to return July 29th. What is the total cost of the return ferry journey for **both** John and James including bicycle stowage?

A   £736
B   £616
C   £592
D   £472

6   The ferry timetable takes into account the difference between British Summer Time, (BST), and The Netherlands Time Zone.

Rotterdam Ferry		
Sailing Timetable	Time of Departure	Sailing Duration
Night crossing	** 21:45	9 hours 45 minutes
Day crossing	** 09:30	7 hours 30 minutes
** NB Arrival time in the Netherlands Time Zone = BST + 1 hour		

John and James prefer to travel on the night crossing. What time are they due to arrive at Rotterdam in the Netherlands Time Zone?

A   07:45
B   08:30
C   09:30
D   09:45

7   John's bank does not charge commission rates on foreign currency exchange transactions. Between them, John and James decide to exchange £300.

> The current exchange rate for Euros to British pounds is:
> €1 = £0.79

Which calculation works out how many Euros they receive for their £300?

A $\dfrac{79}{300 \times 100}$

B $\dfrac{300 \times 100}{79}$

C $\dfrac{79 \times 100}{300}$

D $\dfrac{300}{79 \times 100}$

8   John and James plan to cycle from Rotterdam to a campsite in Den Helder. On the map, the distance between Rotterdam and Den Helder is three and three quarter inches. The scale of the map is 1 inch to 20 miles.

What is the distance from Rotterdam to the campsite in Den Helder?

A   75.00 miles
B   60.75 miles
C   37.50 miles
D   30.75 miles

The distance between Den Helder and Groningen is 70 miles.

> 1 mile = 1.6 kilometres

9   John and James cycle the distance between Den Helder and Groningen at an average speed of 15 kilometres per hour. Approximately how long does it take to cycle between Den Helder and Groningen?

A   4 hours

B $5\frac{1}{2}$ hours

C   6 hours

D $7\frac{1}{2}$ hours

Questions 10 to 13 are about a trainee at Spectrum Painting and Decorating Company.

10 An apprentice is paid the basic rate of £5.50 per hour for 37 hours a week. The overtime rate is the basic hourly rate plus 50%.

During one week he works an extra 4 hours at the overtime rate.

What is the total amount he earns for the week with the additional overtime hours?

A £203.50
B £225.50
C £236.50
D £338.25

11 The apprentice completes vocational training and is awarded a pay rise of 3.5%. Which one of these calculations can he use to work out his new hourly rate?

A £(5.5 × $\frac{3.5}{100}$ + 5.5)

B £$\frac{(3.5 \times 100)}{5.5}$

C £5.5 × $\frac{3.5}{100}$

D £(5.5 + 3.5) × 100

12 The apprentice works out the number of rolls of plain wallpaper he needs to cover the walls in a room. The walls are 2.5 metres high. The wallpaper is 50cm wide and a standard roll of wallpaper is about 10 metres in length. Taking into account the door and the window he estimates he needs to cut 29 lengths to paper the walls.

How many whole rolls of wallpaper will he need to use?

A 5
B 6
C 7
D 8

13 Two of the walls and the ceiling are to be covered with white emulsion paint. Each wall measures 2.5m by 5m. The ceiling measures 4m by 5m. What is the total area to be painted white?

A 50.0m²
B 45.0m²
C 32.5m²
D 25.0m²

Questions 14 and 15 are about a children's nursery.

14 The secretary at Lilliput children's nursery wishes to advertise nursery places in September. She decides to include information on a leaflet and distribute in the local area.

The printer charges fixed fees for design and to set up the printer plus a charge for each leaflet printed:

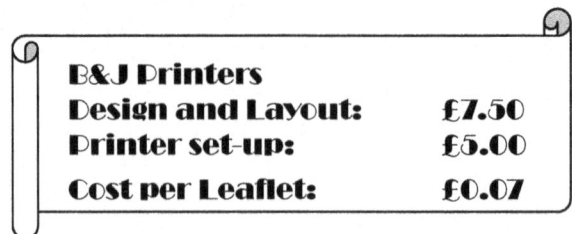

B&J Printers	
Design and Layout:	**£7.50**
Printer set-up:	**£5.00**
Cost per Leaflet:	**£0.07**

What is the cost to design and print 200 leaflets?

A   £12.57
B   £25.14
C   £26.50
D   £37.50

15 The children at the nursery have a mid-morning drink of pure apple juice.

A 750ml bottle of apple juice costs 85 pence and fills 5 x 150ml glasses.

A 3-litre catering pack of apple juice costs £2.10 and fills 20 x 150ml glasses.

How much cheaper, per glass, is the apple juice from the 3-litre catering pack?

A   6.5p
B   7.0p
C   7.5p
D   8.0p

Please go on to the next page

Questions 16 to 21 are about a frozen foods store.

16 The store manager at Freeze-Fresh Foods regularly defrosts and cleans each freezer cabinet before re-stocking with frozen product packets.

The diagram shows the operating temperature of a freezer cabinet.

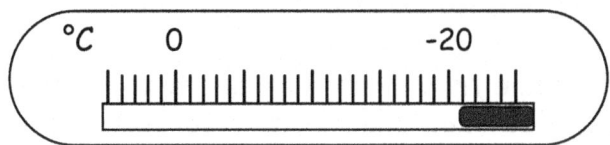

Approximately what is the temperature of the freezer cabinet?

A   - 23°C
B   - 21°C
C   - 19°C
D   - 18°C

17 After defrosting and cleaning, the temperature of the freezer cabinet is 11°C. The temperature must drop to at least -18°C before it can be re-stocked. The temperature of the cabinet must drop by how many degrees?

A   29°C
B   27°C
C   18°C
D   7°C

18 Each freezer cabinet has **three identical sections**. The diagram shows one section of the freezer cabinet with measurements.

Diagram not to scale

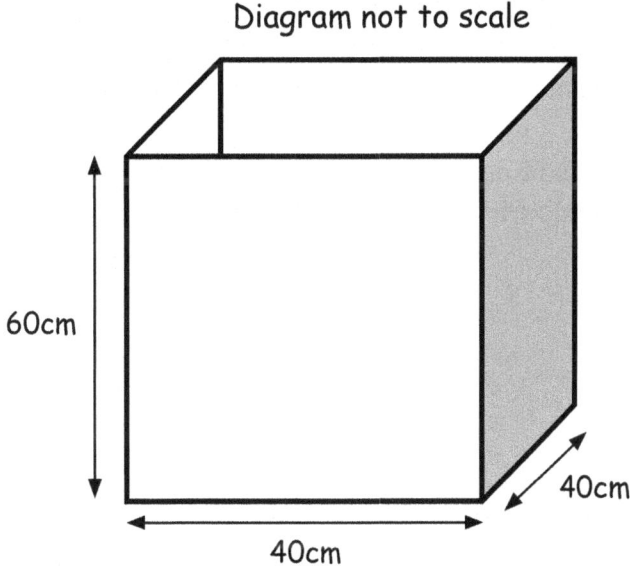

60cm

40cm

40cm

What is the **total** volume, in square centimetres, of the **freezer cabinet**?

A   14000cm³
B   32000cm³
C   96000cm³
D   288000cm³

19   One section of the freezer cabinet will be filled with packets of Cod Steaks which measure 4cm x 10cm x 20cm. What is the maximum number of packets that can fit into **one** freezer cabinet section?

A   360
B   120
C   100
D   80

20   The total sales income from frozen products at Freeze-Fresh Foods in December was £3400. In January the total sales amounted to £1360. Compared with the sales figure for December, sales in January decreased by approximately what fraction?

A $\frac{1}{3}$

B $\frac{2}{5}$

C $\frac{1}{2}$

D $\frac{3}{5}$

The business account in January for this branch of Freeze-Fresh Foods ends with a balance of £1000. The figures for the month are shown in the table. In week 3 the manager buys 2 new freezer cabinets and pays for installation in week 4.

Freeze-Fresh Foods	February Account			
	week 1	week 2	week 3	week 4
Balance carried forward	£1000	£1850	£2952	-£35
Profit / Loss	£850	£1102	-£3005	-£758
Balance at end of week	£1850	£2952	-£35	

21   The balance for the end of February has yet to be included in the table. What is the balance at the end of February?

A   -£793
B   -£723
C   £723
D   £793

Please go on to the next page

Questions 22 to 24 are about the refurbished function room at The Oaks Hotel.

22 The diagram shows a scale drawing of the Oaks Hotel function room.

What is the width, in metres, of the function room?

A   25.0m
B   21.5m
C   17.2m
D   16.0m

23 The Oaks' resident DJ, Beau Jangles, charges £6.50 per hour and a fixed fee of £75 to play music for parties in the function room. What is the total cost to hire Beau Jangles for a party starting at 7:30pm until 1am?

A   £81.50
B   £107.50
C   £110.75
D   £120.50

24 To comply with fire regulations, the site foreman of the refurbishment must calculate the maximum number of persons that the function room can accommodate.

Each person must have a minimum of 0.45 metres of floor space.

| L = length in metres; W = width in metres |

Which of the formulas will the site foreman use?

A $\dfrac{L \times W}{0.45}$

B $\dfrac{L + W}{0.45}$

C $\dfrac{0.45}{L \times W}$

D $\dfrac{0.45}{L + W}$

Questions 25 to 27 are about a maker of novelty cakes.

25 An order for a large birthday cake, in the shape of a football field, is to be made up of a number of smaller cake squares with a layer of green icing on the top.

The basic recipe shows the amounts of each ingredient for one small square that will be baked in a tin measuring 15cm x 15cm.

> *Basic Sponge Cake Recipe*
>
> *125g Self-Raising Flour*
> *125g Caster Sugar*
> *125g Polyunsaturated Margarine*
> *3 Large Free-Range Eggs*

The football field cake will measure 45cm x 60cm. How much self-raising flour is needed to make the large birthday cake?

A   1.25 kilograms
B   1.50 kilograms
C   1.75 kilograms
D   1.80 kilograms

26 A layer of green icing for the football field cake measures 45cm x 60cm x 1.5cm. How many cubic centimetres of icing are used to cover the cake?

A   2700cm³
B   4050cm³
C   5400cm³
D   6075cm³

27 A cake in the shape of a football is topped with chocolate icing.

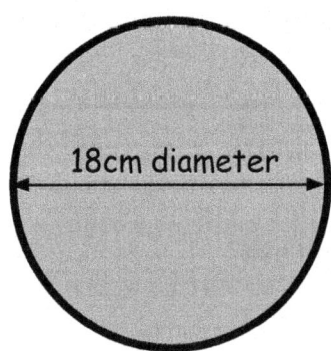

18cm diameter

Area of a circle = π x r²
(π = 3; r = radius)

Diagram not to scale

What is the approximate area of the top of the football shaped cake?

A   243cm²
B   360cm²
C   486cm²
D   972cm²

Questions 28 and 29 are about a commercial property agent.

28 A letting agent uses a bar chart to show the rise and fall of commercial property prices in Southumberland in the past 7 years.

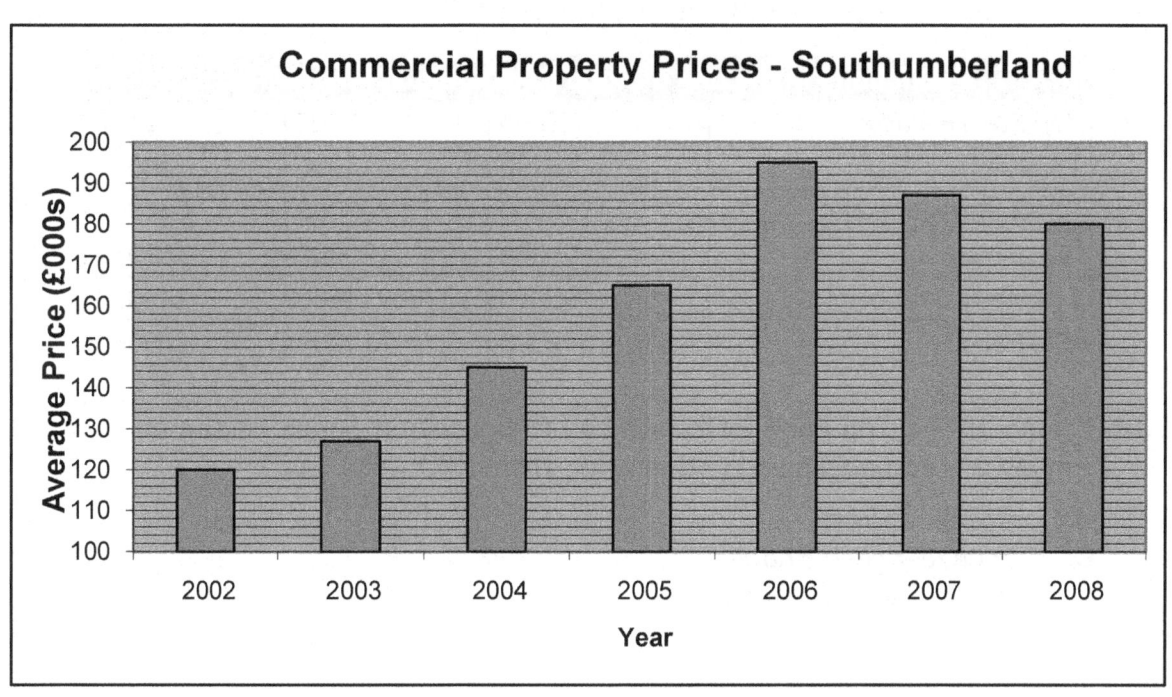

In what way is this chart misleading?

A The scale on the vertical y-axis has gaps of one hundred
B The scale on the vertical y-axis starts at a number above zero
C The scale on the horizontal x-axis is irregularly spaced
D The scale on the horizontal x-axis starts at a number above zero

29 A factory unit on an irregularly shaped plot of land requires a security fence and gates to be erected for the property agent as a condition of sale.

The diagram shows the dimensions of the plot of land to be fenced off.

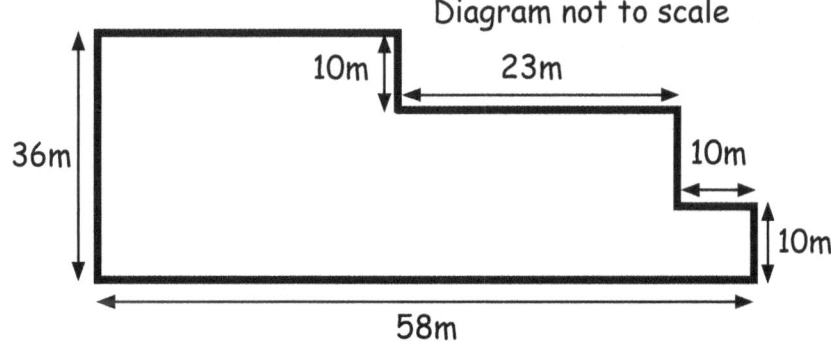

What is the total length of the perimeter of the plot of land?

A 147m
B 163m
C 172m
D 188m

Questions 30 to 32 are about the efforts of supermarket chains to reduce the amount of food packaging used in their products.

The table compares the reduction in food packaging, in tonnes, made by the top five food retailers in the Northeast Region from 2005 to 2007.

Reductions in the use of Food Packaging (Tonnes/Year)			
Top Food Retailers	2005	2006	2007
Freshco	13000	14000	16500
Allied Supermarkets	305000	450000	561000
Greenmarket	95000	170000	257000
Freeze-Fresh Foods	195000	214000	249000
Vita Foods	649000	712000	709000

30 Which of the Top Food Retailers has the greatest range, in the number of tonnes of food packaging reductions, from 2005 to 2007?

A   Greenmarket
B   Freeze-Fresh Foods
C   Allied Supermarkets
D   Vita Foods

31 Business Administration student, Anna, considers methods to display the **trends** in the reductions of food packaging made by retailers from 2005 to 2007.

What type of representation should Anna use?

A   a comparative bar chart
B   a comparative line graph
C   a pie chart
D   a pictogram

32 Vita Foods have stores all over the UK. The total reduction in food packaging for the whole company in 2007 was 3,500,000 tonnes.

Approximately what fraction of the total reduction in food packaging in 2007 was made by Vita Foods stores in the Northeast region?

A $\frac{1}{8}$

B $\frac{1}{7}$

C $\frac{1}{5}$

D $\frac{2}{7}$

Questions 33 to 36 are about calls to a commercial call-centre.

A call-centre section handles inbound business calls for three separate accounts:

Business Account Name	Response Time Target (RT)	Service Level
X – Post Code Directory	20 seconds	4
Y – T.V. Licensing	05 seconds	1
Z – Post Service Directory	15 seconds	2

Businesses pay extra bonuses in their service level contract to have their calls answered within a set response time.

The table shows response times (in seconds) by one operator handling 30 inbound calls:

Call No:	1	2	3	4	5	6	7	8	9	10	RT	% Achieved
X	12	18	17	13	21	23	16	14	20	11	20	
Y	02	08	02	05	02	08	03	06	02	01	05	100
Z	14	15	15	14	15	14	14	15	14	15	15	100

33   What was the percentage of calls for account **X** that met the target response time?

   A   30%
   B   75%
   C   80%
   D   85%

34   What is the mean response time for account **Z**?

   A   14.0 seconds
   B   14.5 seconds
   C   15.0 seconds
   D   15.5 seconds

35   What is the modal response time for account **Y**?

   A   2 seconds
   B   3 seconds
   C   5 seconds
   D   8 seconds

36   What is the range of response times for accounts **X**, **Y** and **Z**?

   A   12 seconds
   B   15 seconds
   C   22 seconds
   D   23 seconds

Questions 37 and 38 are about the leisure pursuit of angling.

37   Two fishermen are awarded prizes for their entries in a fishing competition.

The fish are weighed to decide the winner for the heaviest fish caught:

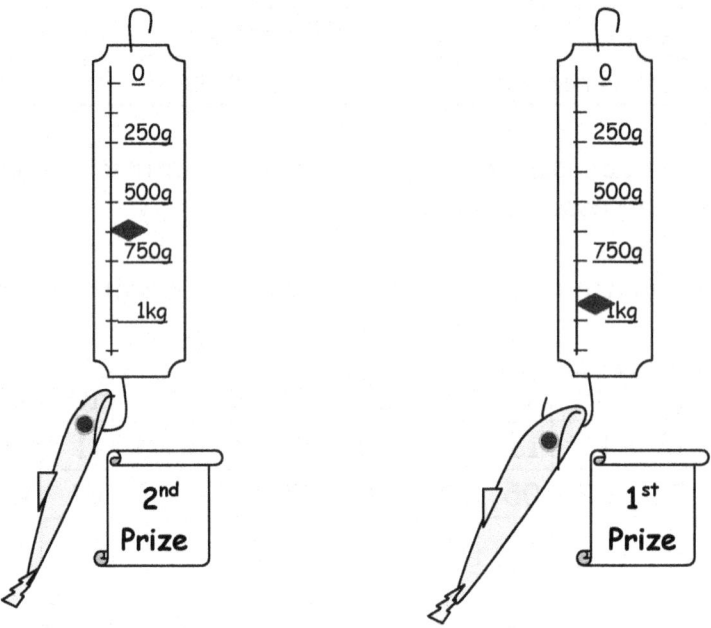

37   How much heavier is the 1st prize winner than the runner-up?

A   625.0g
B   312.5g
C   250.0g
D   212.5g

38   Mealworms are used as bait to attract fish to the angler's fishing line. A quantity of mealworms is placed on the weigh scale.

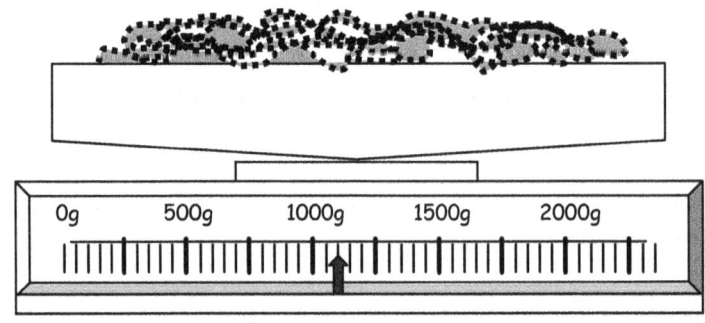

What is the approximate weight of the mealworms on the scale, in kilograms?

A   1.10kg
B   1.19kg
C   1.21kg
D   1.20kg

Questions 39 and 40 are about traffic on UK roads.

39   The chart shows the proportion of car journeys on road types in the UK.

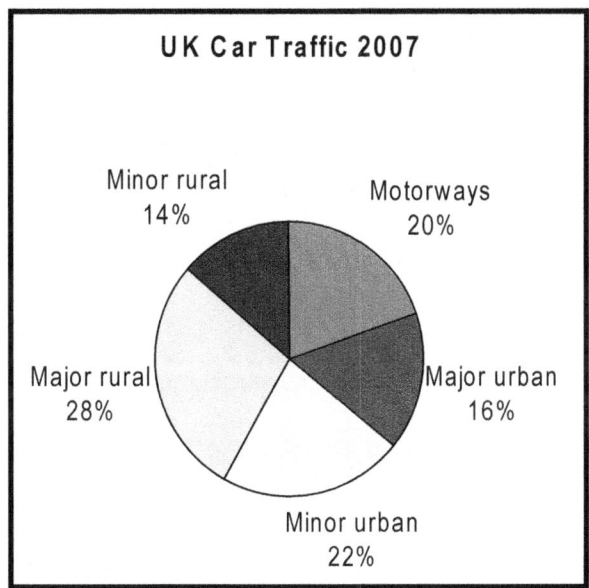

Referring to the chart, which of these statements is correct?

A   more car journeys are made on urban roads than on rural roads
B   more car journeys are made on rural roads than on urban roads
C   most car journeys are made on motorways and major urban roads
D   most car journeys are made on motorways and minor rural roads

40   The graph shows the trend in petroleum consumption by UK transport.

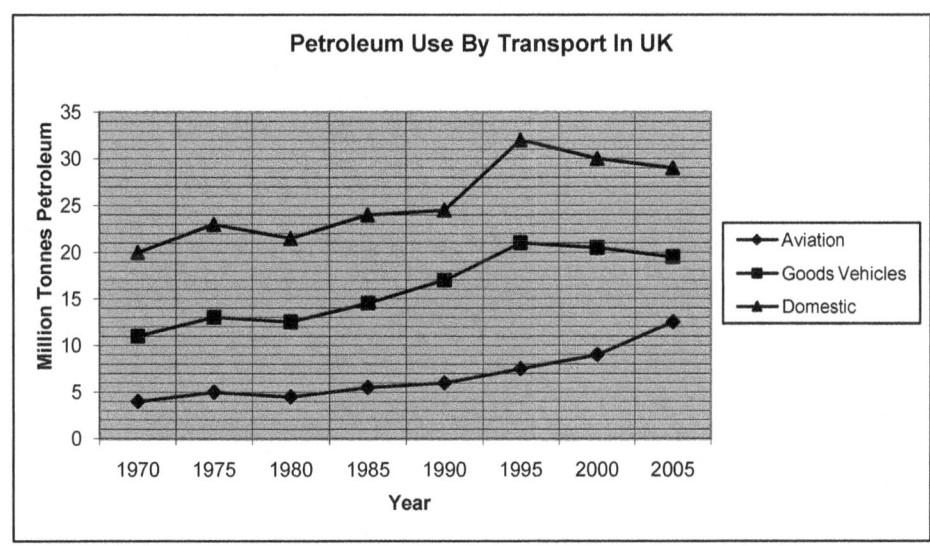

Which statement accurately describes petrol consumption as illustrated in the graph?

A   petrol consumption by Aviation is on a continuous downward trend
B   petrol consumption by Goods Vehicles is on continuous upward trend
C   petrol consumption by Aviation and Goods Vehicles show a downward trend since 1995
D   petrol consumption by Domestic transport reached a peak in 1995

End of Paper

Practice Multiple-choice Paper
Suitable for:

Key Skills Level 2 Application of Number
Level 2 Adult Numeracy

Paper Ten

YOU NEED

- ■ This test paper.
- ■ A pen.
- ■ A pencil and eraser.
- ■ An Answer Sheet.
- ■ A ruler marked in centimetres and millimetres

You may NOT use a calculator.
You may use a bilingual dictionary.
There are 40 questions on this paper. Try to answer ALL the questions.
When you have completed the questions you must check your answers, then check them again.

YOU HAVE QUARTER OF AN HOUR TO READ THE PAPER
AND ONE HOUR TO COMPLETE THE 40 QUESTIONS

INSTRUCTIONS

- ■ Make sure you write your name and today's date on the Answer Sheet. Use a pen to do this.
- ■ Use a pencil to mark your answers so if you change your mind you can erase your choice and select another.
- ■ Make sure that for each question you have only selected one answer. If you select more than one, the answer will not be marked.
- ■ Read each question carefully before you select an answer.

Note for learners and tutors: This is a practice test that has been designed to closely resemble the questions and question styles of a "live" paper.

Questions 1 to 3 are about a train journey to the UK's capital city.

Brothers, Bill and Phil, plan to travel to visit their sister, Jill, in London. They use the timetable below to work out travel arrangements. They plan to catch the latest possible train from Sunderland to arrive in London before mid-day.

Newcastle – London	Departure times			
Newcastle	0552	0652	0822	1022
Sunderland	0620	0720	0830	1030
Hartlepool	0635	0735	0845	1045
Eaglescliffe	0655	---	0905	1105
Northallerton	0715	0800	0925	1125
Thirsk	0755	0840	1005	1205
York	0825	0910	1035	1235
London	1025	1110	1235	1435

1   Bill and Phil live within a 20-minute taxi ride away from the train station in Sunderland and they allow an extra 15 minutes in case of traffic hold-ups. What time must they leave the house to catch the train that will get them to London before mid-day?

A   6:20am
B   6:45am
C   6:52am
D   7:00am

2   The train arrives on time. Bill and Phil are met by their sister Jill at the train station. Twenty-five minutes later they arrive at Jill's house. How many hours and minutes does it take for Bill and Phil to travel from the station in Sunderland to Jill's house?

A   4 hours 15 minutes
B   4 hours 18 minutes
C   4 hours 33 minutes
D   4 hours 43 minutes

3   The standard return train fare from Sunderland to London is £96. As Bill and Phil booked four weeks in advance their saver tickets cost 35% less than the standard fare. How much did each saver ticket cost?

A   £61.00
B   £62.40
C   £65.00
D   £92.50

Please go on to the next page

Questions 4 to 7 are about wholesale and retail coffee sales.

Carrington's Coffee Merchants supply a range of coffee beans to shops and catering businesses in the UK.

Coffee Sales – Ethically Traded Varieties (million kilos per year)		
Coffee Bean Variety	2006	2007
Columbian	1.03	1.16
Brazilian	0.42	0.57
Puerto Rican	0.41	0.47
Kenyan	0.31	0.31
Arabica	0.23	0.25
Total UK Sales Ethically Traded Varieties	2.40	2.76
Total UK Sales All Varieties	17.00	19.32

4   What weight of Carrington's Columbian coffee beans was sold in 2006?

   A   10 300 kilograms
   B   103 000 kilograms
   C   1 030 000 kilograms
   D   10 300 000 kilograms

5   What is the percentage increase in the total sales of all Ethically Traded Varieties of coffee from 2006 to 2007?

   A   15.0%
   B   12.0%
   C   07.5%
   D   01.5%

6   In 2007, what fraction of the Total UK Sales of All Varieties of coffee is the Total UK Sales of Ethically Traded Varieties in that year?

   A $\frac{1}{5}$

   B $\frac{1}{6}$

   C $\frac{1}{7}$

   D $\frac{1}{8}$

Please go on to the next page

7   The logistics controller for Carrington's Coffee Merchants works out the lowest mileage route for deliveries to three towns from the local depot in Carlton.

The diagram shows the miles between the depot and the delivery locations:

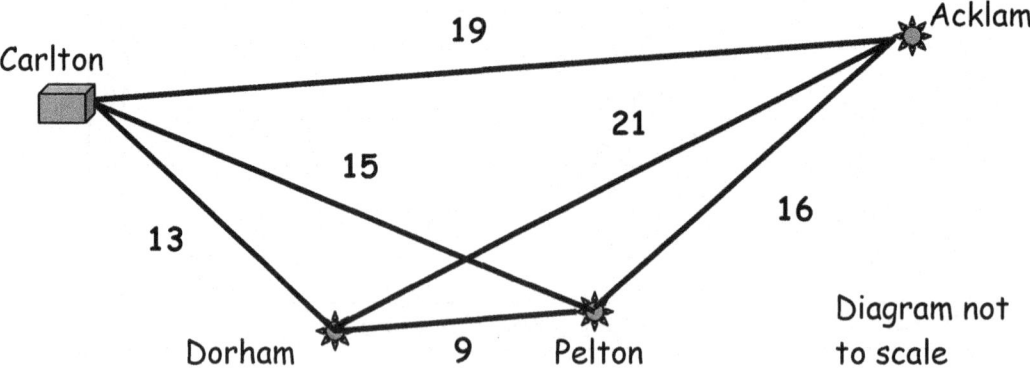

Diagram not to scale

Which route involves the least mileage?

A   Carlton – Dorham – Acklam – Pelton - Carlton
B   Carlton – Dorham - Pelton - Acklam - Carlton
C   Carlton – Acklam – Dorham – Pelton - Carlton
D   Carlton – Pelton – Acklam – Dorham - Carlton

Questions 8 to 12 are about Foresters Woodland Adventure Park.

8   A family group comprising 4 adults and 6 children plan to hire a holiday let for two weeks near Foresters Woodland Adventure Park. They have a budget of no more than £700 for accommodation and want to be within 10 miles of the Park.

Details of a number of possible holiday lets are shown in the table.

Holiday Home	Miles from Foresters Park	Weekly Hire	Maximum number of persons
Cragg Lea	10	£425	12
Mere Vista	16	£370	12
Lith-na-gow	9	£340	8
Stable Villa	10	£326	10
Mount Holme	15	£324	8
River View	5	£349	6

Which of the holiday homes meets the needs of this family group?

A   Mere Vista
B   Lith-na-gow
C   Stable Villa
D   River View

9   A group of 4 adults and 6 children (each child under 18 years of age) plan a day's visit to Foresters Woodland Adventure Park. The table shows the cost of day passes.

Day Pass Monday to Friday		Day Pass Saturday and Sunday	
Adult	Child*	Adult	Child*
£7.50	£3.00	£9.50	£5.00
* up to 18 years			
NB - Discount of 10% for parties of ten or more Monday to Wednesday.			

How much **more expensive** is it for this group to visit the Park on Saturday or Sunday than it is on Monday, Tuesday or Wednesday?

A   £30.00
B   £24.80
C   £20.00
D   £16.80

10   In total 65,011 visitors enjoyed Foresters Woodland Adventure Park last year. 19,980 of the total number of visitors were adults.   What was the approximate ratio of adults to children?

A   4 : 9
B   9 : 4
C   1 : 3
D   2 : 5

Please go on to the next page

11 Foresters Woodland Adventure Park sponsors cross-country cycling races. The graph shows the progress of the lead cyclist during Stage 1 of a race.

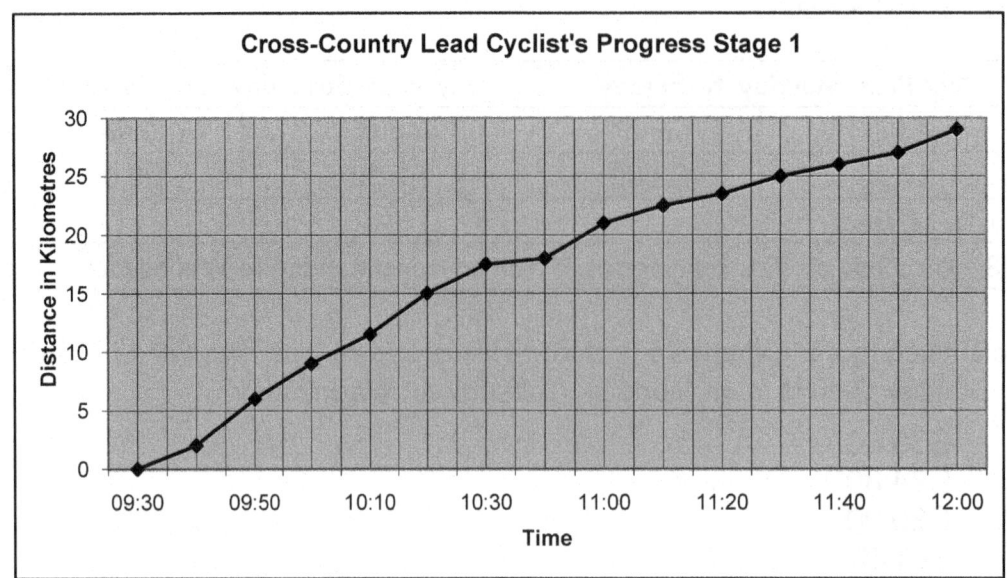

How far does the lead cyclist travel between 10:00am and 11:00am?

A 13 kilometres
B 12 kilometres
C 11 kilometres
D 10 kilometres

The end of Stage 2 of the race, shown in the graph, is reached at 18:00.

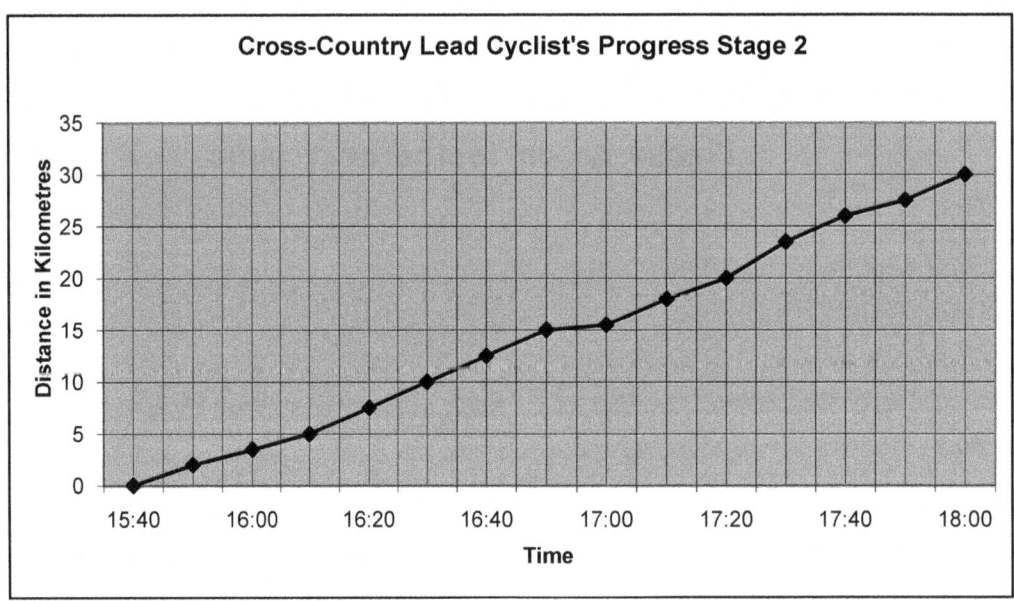

12 How long does it take the lead cyclist to travel the last 15 kilometres?

A 1 hour 30 minutes
B 1 hour 20 minutes
C 1 hour 10 minutes
D 50 minutes

Questions 13 to 15 are about commercial activities in a fresh-water fish farm.

13 The water temperature in two trout hatcheries is monitored over a two-week period and compared in a line graph.

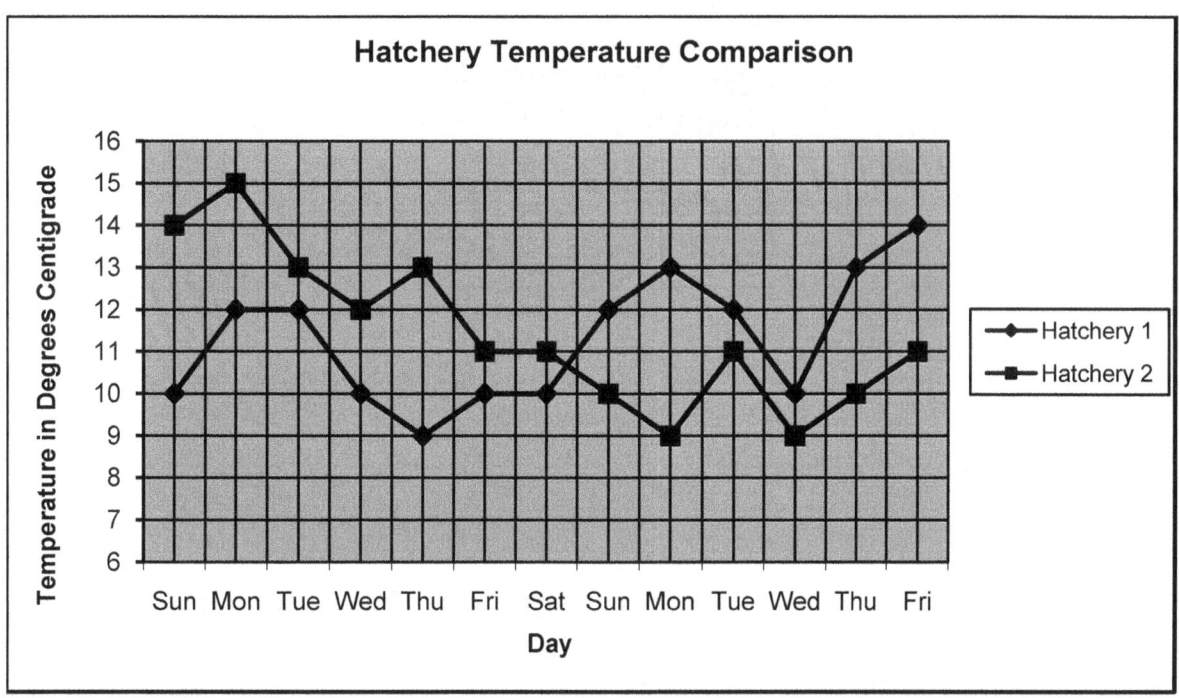

During the two-week period, on how many days was the temperature in Hatchery 1 **more than two degrees** centigrade higher than in Hatchery 2?

A   6 days
B   5 days
C   4 days
D   3 days

14 In the regional area where the fish farm is located, the mean monthly rainfall is 65 millimetres. Records for 2006 and 2007 are shown in the table.

Recorded Monthly Rainfall (millimetres)												
Year/Month	Jan	Feb	Mar	Apr	May	Jun	Jul	Aug	Sep	Oct	Nov	Dec
2006	56	71	86	69	55	60	44	40	32	111	82	86
2007	83	64	102	57	19	77	57	61	34	82	79	42

In how many months in the two-year period was the **recorded** monthly rainfall greater, by at least 20 millimetres, than the **mean** monthly rainfall?

A   4
B   5
C   7
D   8

15 The fish farm employs hourly paid workers to pack orders ready for delivery. Packers are paid: £6.60 per hour, Monday to Friday; £8 per hour at weekends.

The hours worked by one packer are shown in the table:

Day	Hours worked
Thurs	7.5
Fri	7.5
Sat	6.5
Sun	4.0

How much pay did the packer earn in total?

A £184
B £183
C £100
D £99

Questions 16 to 19 are about business activities of a charity organisation.

16 A national poll surveyed UK companies to ask how many charities they supported.

Charities Supported		
No of Charities	Private Sector	Public Sector
0	4524	5887
1	3171	2683
2	6709	1934
3	221	806

How many companies in both the Private Sector and the Public Sector supported two or more charities?

A 9670
B 9449
C 8643
D 8422

Please go on to the next page

17 The chart shows the thousands of stock items sold by a charity shop in two successive years.

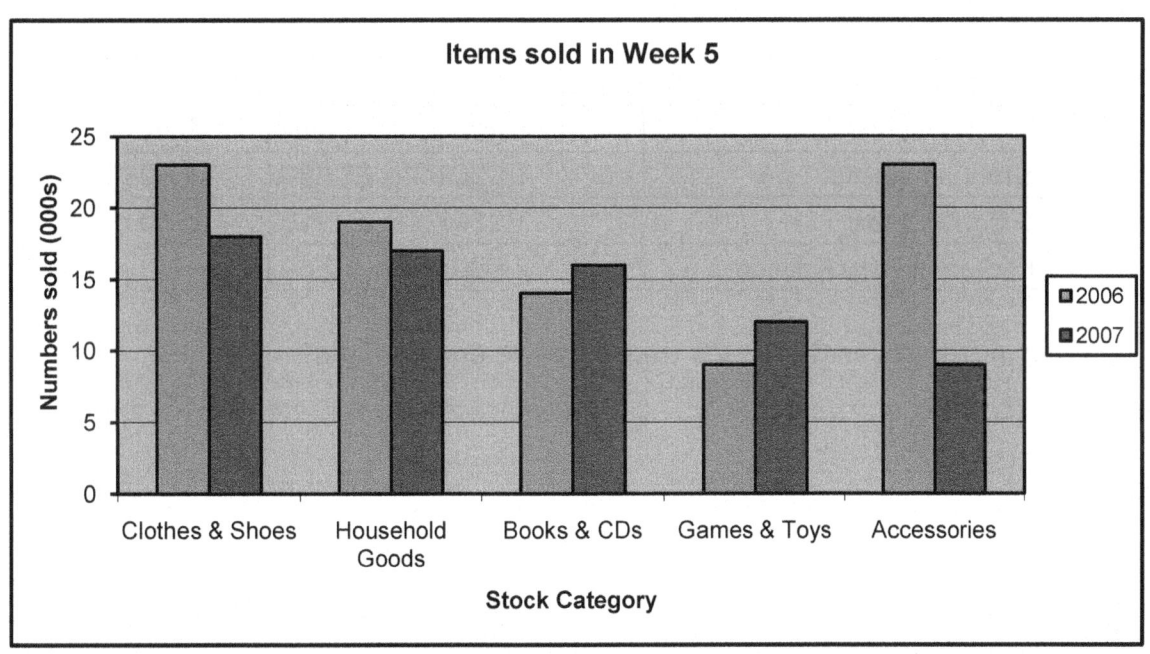

How many **more** Books & CDs and Games & Toys were sold in 2007 than in 2006?

A   1000
B   4000
C   5000
D   6000

18 The budget for a charity shop is shown in the table.

Shop Budget		
Expenditure	**2006**	**2007**
Staff wages	170500	174500
Pension contributions	800	840
Payroll / Personnel administration	8000	8200
Staff Training – course fees	2150	2250
Staff Training – travel expenses	1500	1730
Stock transport expenses	3010	3560
Insurances	1850	1950

What is the difference between **total** staff training expenditure for 2006 and 2007?

A   £315
B   £330
C   £220
D   £100

19   The charity shop begins a new financial year showing a balance of £950 in credit. The table shows the credit / debit balance sheet for three months. The balance at the end of June has not yet been calculated.

Balance Sheet			
	April	May	June
Starting Balance	£950	£1,880	-£60
Credit / Debit	£930	-£1940	-£1360
End-of-month Balance	£1,880	-£60	

What is the correct balance figure at the end of June?

A   £1300
B   -£1300
C   £1420
D   -£1420

Questions 20 to 22 are about an industrial cleaning company, Kleanz-Co.

20   A business hires Kleanz-Co to clean the carpets in their office complex from Thursday to Sunday.

Kleanz-Co - Carpet cleaning service charges		
	Standard Service	Premium Service
One Day	£165	£185
Extra Days	£130	£150
One week (7 days)	£750	£850
Weekend	£210	£230

Using the Premium Service, how much will the business pay for the carpet cleaning service on Thursday, Friday and the weekend?

A   £500
B   £565
C   £750
D   £850

Please go on to the next page

21   The hours worked, by a Kleanz-Co employee, are shown in the table:

Day	Shift	Hours
Wednesday	Day	3.5
Thursday	Night shift	7.5
Friday	Night shift	7.5
Saturday	Over time	7.5
Sunday	Over time	7.5

Rates of pay are:
Monday to Friday, Day shift   £7.60 per hour
Monday to Friday, Night shift   £10.00 per hour
Saturday and Sunday Over time   £12.50 per hour

How much does the Kleanz-Co employee earn in the week?

A   £150.00
B   £187.50
C   £213.80
D   £364.10

22   The Kleanz-Co employee earns an average of £1200 per month. The pie chart shows what he spends his salary on:

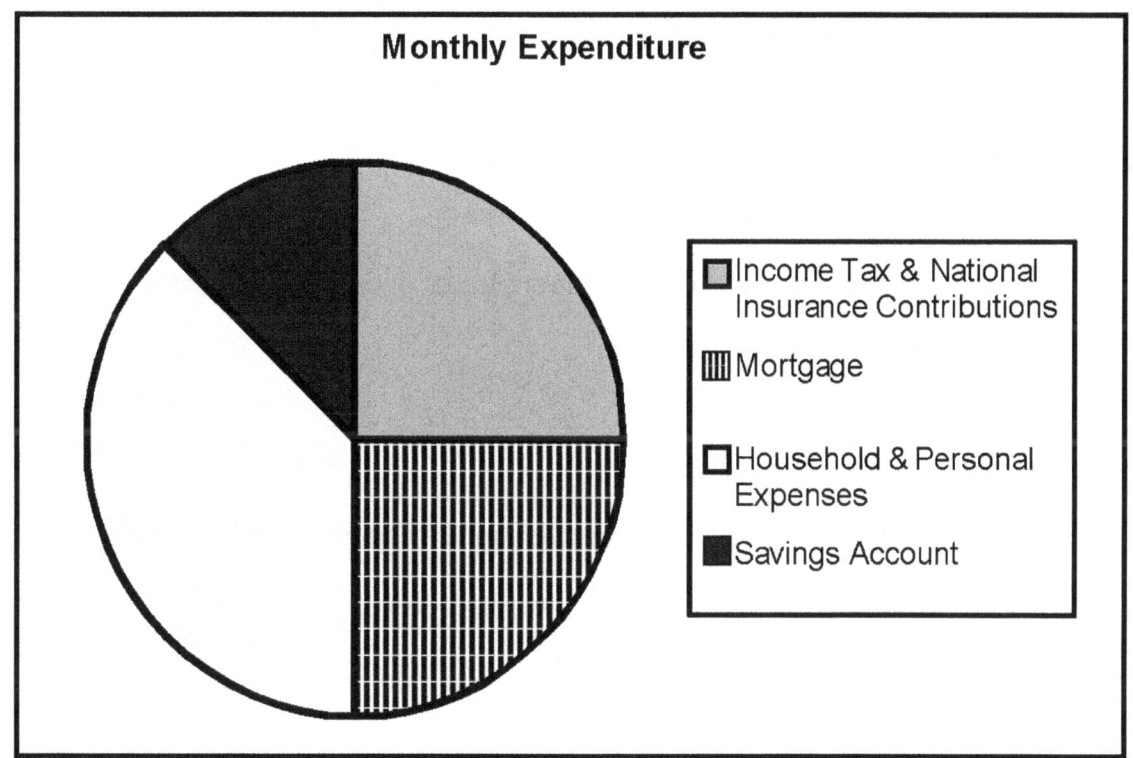

Approximately what fraction of his income goes into his savings account?

A   1/6
B   1/8
C   1/10
D   1/12

Questions 23 to 27 are about temperatures in a market garden greenhouse.

The table shows the overnight temperatures in °C recorded in one market garden greenhouse during a **two-week** period.

	Monday	Tuesday	Wednesday	Thursday	Friday	Saturday	Sunday
Week 1	8	10	13	8	9	6	5
Week 2	8	11	12	11	8	7	3

23   What is the mean of the overnight temperatures?

    A   7.0°C
    B   7.5°C
    C   8.0°C
    D   8.5°C

24   What is the median overnight temperature?

    A   6°C
    B   7°C
    C   8°C
    D   9°C

25   What is the modal number of the overnight temperatures?

    A   8°C
    B   7°C
    C   4°C
    D   3°C

26   What is the range of overnight temperatures over the two-week period?

    A   11°C
    B   10°C
    C   9°C
    D   8°C

27   For what fraction of days was the overnight temperature less than 9°C?

    A   4/7
    B   3/7
    C   2/7
    D   1/7

Please go on to the next page

Questions 28 and 29 are about service activities in a call centre.

28 A team bonus is paid when a certain percentage of all Service Agreement targets are achieved. The percentage rates for bonus payments for one call centre team are as follows:

Charity Pledges: 75%
Post Sorting Service: 80%
T.V. Licensing: 90%

The table shows the latest performance monitoring information:

Service Agreement - Call Handling Response Times				
Call Handling Account	Target Response Time (seconds)	Calls Taken	Mean Response Time (nearest second)	% Target Achievement
Charity Pledges	15	21	5	92%
Post Sorting Service	10	64	12	72%
T.V. Licensing	6	33	14	100%

Which statement is true? To achieve a bonus, the team must:

A   increase the % of Charity Pledge calls answered within 15 seconds
B   reduce the % of Charity Pledge calls answered within 15 seconds
C   increase the % of Post Sorting Service calls answered within 10 seconds
D   reduce the % of Post Sorting Service calls answered within 10 seconds

29 In the first hour of a televised charity fundraising event, a call centre operative receives calls pledging the following amounts:

19 viewers pledge £20 each
23 viewers pledge £10 each
12 viewers pledge £5 each

Which calculation works out the total, in pounds, pledged by viewers in the first hour?

A   (19 + 23 + 12) × (20 + 10 + 5)
B   (19 × 20) + (23 × 10) + (12 × 5)
C   (19 × 20 + 23 × 10 + 12 × 5) ÷ 100
D   (19 × 20 + 23 × 10 + 12 × 5) × 100

Please go on to the next page

Questions 30 to 33 are about the Marine and Technical College.

30 The Marine and Technical College publishes an information booklet including details of staff numbers in each department. The table shows changes in staff numbers from 2005-06 to 2006-07.

Department	Staff Numbers 2006-07	Change from 2005-06
Beauty & Hair	27	+2
Business	29	+4
Catering & Hospitality	14	+3
Education & Training	22	-6
Horticulture	10	+1
Leisure & Tourism	11	-3
Marine Technologies	78	+10
Motor Vehicle	15	+2
Support Staff	25	-21
Technical Staff	22	-3
Uniformed Services	7	-2

What is the change in the total number of college staff from 2005-06 to 2006-07?

A   57 more staff members
B   57 fewer staff members
C   13 more staff members
D   13 fewer staff members

A group of Marine Technologies apprentices go on a field trip to Sigma oilrig in the North Sea. The line on the scale drawing indicates the rig's distance from the mainland.

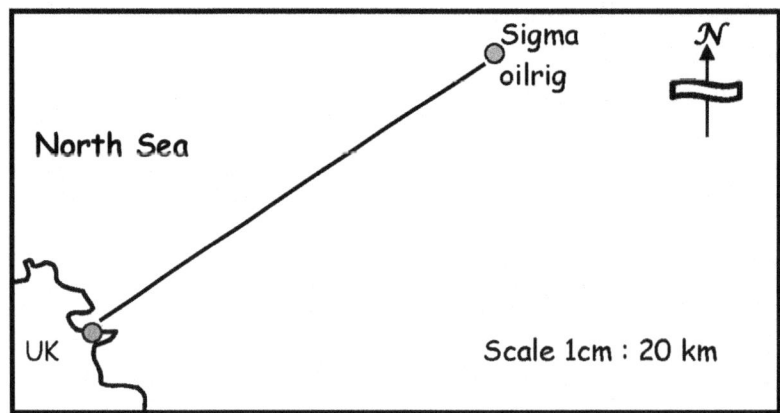

31 What is the actual distance, in kilometres, from the mainland to the Sigma oilrig?

A   120 kilometres
B   126 kilometres
C   175 kilometres
D   206 kilometres

Apprentices in the Marine Technologies welding laboratory use oxy-acetylene gas stored in a metal cylinder. The gas cylinder is fitted with a pressure gauge to indicate the units of gas it contains. The diagram shows the gauge before and after a welding workshop.

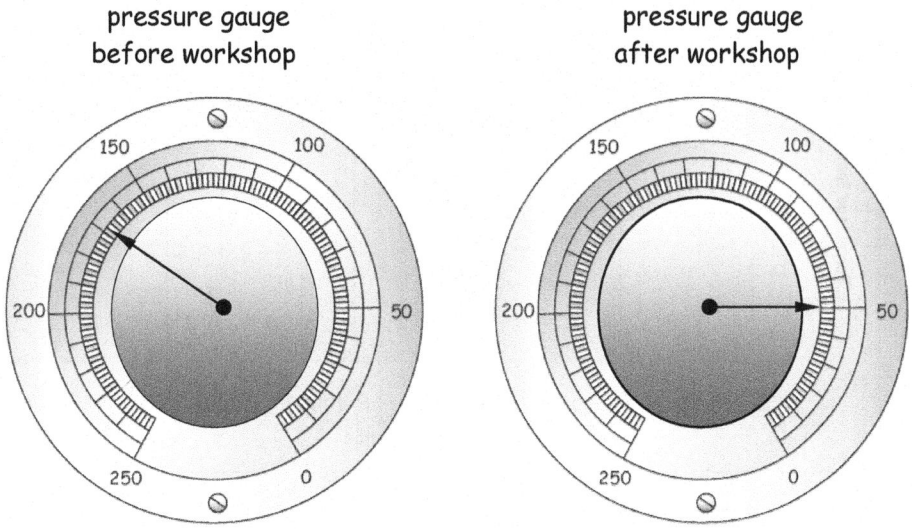

pressure gauge
before workshop

pressure gauge
after workshop

32   To the nearest 10 units, how many units of gas have been used in the period before and during the workshop?

   A   170 units
   B   168 units
   C   120 units
   D   118 units

Please go on to the next page

33 The chart shows maths test percentage scores for 7 motor vehicle apprentices:

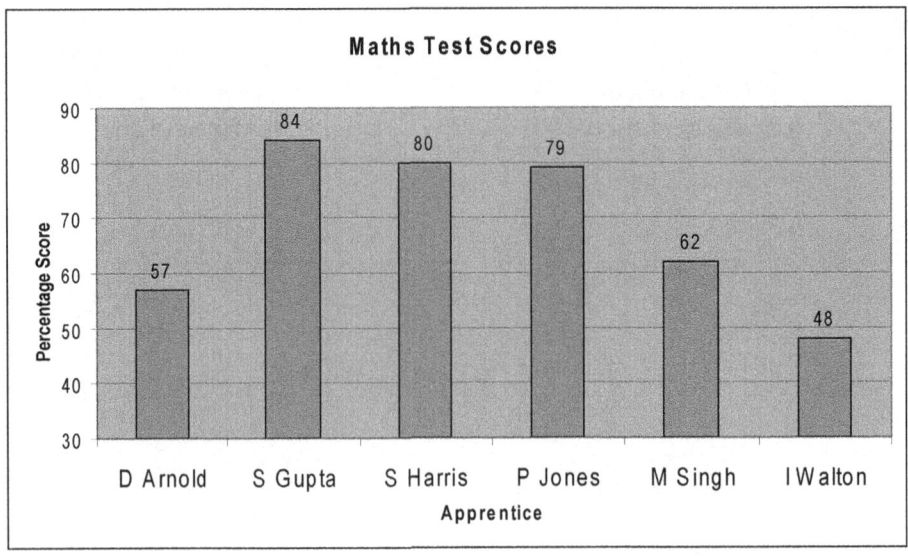

In which way is the chart misleading?

A   the x-axis has uneven gaps between the bars
B   the y-axis does not start at zero
C   the scores for the test are not out of 100
D   the horizontal grid lines are too far apart

Questions 34 and 35 are about the expansion of a garden nursery business.

The gardener decides to increase greenhouse space on a spare plot of land. The shape and dimensions of the new greenhouse space are shown in the diagram.

Diagram not to scale

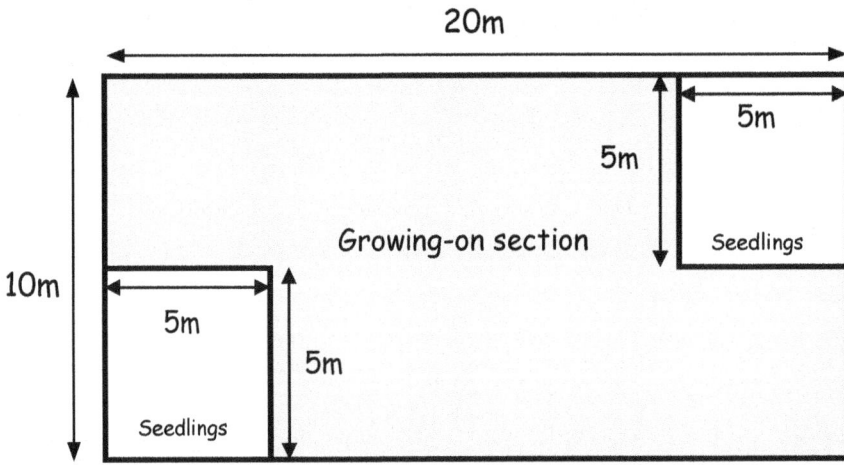

34 A landscape contractor charges £10 per square metre to construct a concrete floor. How much does it cost to construct the greenhouse floor, measuring 20m x 10m?

A   £200
B   £300
C   £2000
D   £3000

35 In the new greenhouse there are 2 areas, each measuring 5m x 5m, where the seeds are planted. The rest of the greenhouse space is the Growing-on section where young plants grow to their mature height.

What is the area, in square metres, of the Growing-on section?

A 50m²
B 75m²
C 100m²
D 150m²

Questions 36 to 38 are about preparing a product for the European market.

36 Allied Foods make two sizes of traditional British recipe fruitcake for export to European customers. The small size cake weighs 2 pounds and 8 ounces (2lb 8oz).

1 pound (1lb) is 16 ounces (oz)	1oz is approximately 28g

What is the approximate weight of the small fruitcake in kilograms?

A 1.12kg
B 1.23kg
C 1.34kg
D 1.45kg

37 The weight and volume of the fruitcake are used to work out transport costs.

To calculate the volume of the cake, multiply the surface area by the depth.

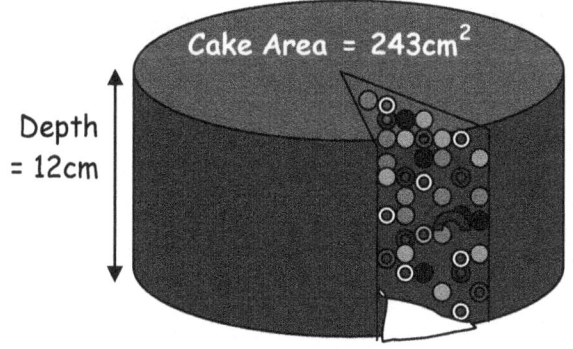

Cake Area = 243cm²

Depth = 12cm

Volume = Area x Depth

Diagram not to scale

What is the approximate volume of a whole small fruitcake, in cubic centimetres?

A 2430cm³
B 2550cm³
C 2916cm³
D 2956cm³

38   The large size cake sells for £4.50 in the UK. Ten percent is added to the UK price to cover packaging and export costs.

$$£1 = €1.26$$

What is the cost of the cake for the European Market in Euros, rounded to 2 decimal places?

A   €6.67
B   €6.24
C   €5.67
D   €4.54

Questions 39 and 40 are about the manufacture of a bronze sculpture.

39   A sculptor makes a scale model of a bronze figure. The model is 15cm tall. The full size sculpture will be 3.0 metres tall.

To what scale is the model made in comparison to the full size bronze sculpture?

A   1 : 5
B   1 : 10
C   1 : 15
D   1 : 20

40   The full-scale bronze sculpture weighs 300 kilograms. Bronze is made from copper and tin in the following proportions:

Bronze = 91 parts Copper to 9 parts Tin

What weight of Tin is used in the full-scale bronze sculpture?

A   2.7 kilograms
B   3.0 kilograms
C   27 kilograms
D   30 kilograms

End of Paper

Practice Multiple-choice Paper
Suitable for:

Key Skills Level 2 Application of Number
Level 2 Adult Numeracy

Paper Eleven

YOU NEED

- ■ This test paper.
- ■ A pen.
- ■ A pencil and eraser.
- ■ An Answer Sheet.
- ■ A ruler marked in centimetres and millimetres

You may NOT use a calculator.
You may use a bilingual dictionary.
There are 40 questions on this paper. Try to answer ALL the questions.
When you have completed the questions you must check your answers, then check them again.

YOU HAVE QUARTER OF AN HOUR TO READ THE PAPER
AND ONE HOUR TO COMPLETE THE 40 QUESTIONS

INSTRUCTIONS

- ■ Make sure you write your name and today's date on the Answer Sheet. Use a pen to do this.

- ■ Use a pencil to mark your answers so if you change your mind you can erase your choice and select another.

- ■ Make sure that for each question you have only selected one answer. If you select more than one, the answer will not be marked.

- ■ Read each question carefully before you select an answer.

Note for learners and tutors: This is a practice test that has been designed to closely resemble the questions and question styles of a "live" paper.

Questions 1 to 6 are about activities at Silkstone Ski Sport Centre.

1   Silkstone Ski Sport Centre has an artificial ski-slope and ski lift for ski sport practice. Evening bookings are available as the ski-slope has flood lighting.

From Monday to Sunday, the flood lighting is switched on 1 hour and 20 minutes before sunset and switched off when the centre closes at 20:45.

The table shows the sunset times for one week in November.

Day	Sunset time
Monday	16:08
Tuesday	16:06
Wednesday	16:04
Thursday	16:03
Friday	16:01
Saturday	15:59
Sunday	15:58

On Sunday evening for how long, in hours and minutes, is the flood lighting on?

A   4 hours 47 minutes
B   5 hours 07 minutes
C   5 hours 47 minutes
D   6 hours 07 minutes

2   The chart shows the cost of electricity per day depending to the number of hours per day that the flood lighting is switched on. In February, the flood lighting is used for four hours per day at the High power level.

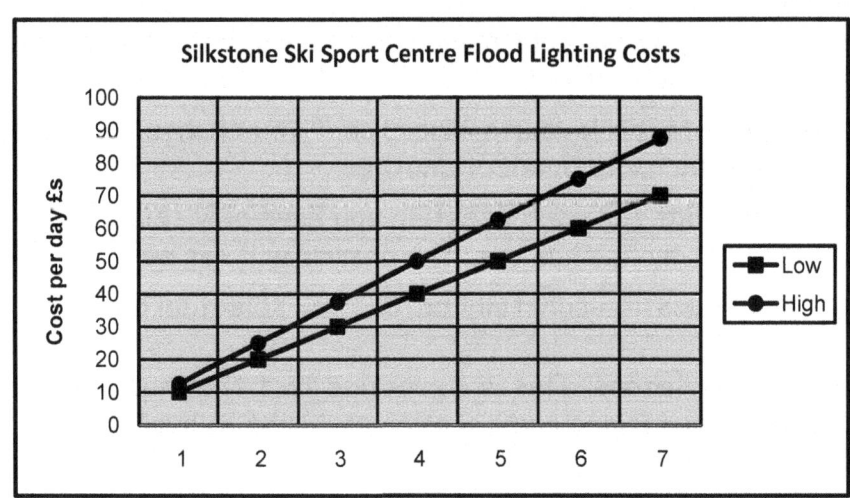

Reading from the graph, what is the approximate cost of lighting the ski-slope for seven days in February on High power?

A   £270
B   £280
C   £350
D   £360

3   The Silkstone Ski Sport Centre membership club offers members 10% discount on holiday bookings to selected European ski-resorts.

Silkstone Ski Club 4-day Ski Break in February (Tuesday – Friday)			
Airport	Ski Resort	Cost per Person	Discount 10%
Berne	Interlaken	£189	Yes
	Wengen	£170	No
Geneva	Chamonix	£185	Yes
	Les Crosets	£165	No
Munich	Garmisch-Partenkirchen	£190	Yes
	Schliersee	£169	No
Treviso	Asagio	£195	Yes
	Pago	£179	No

A family of four book a four-day Ski Break in Chamonix. What is the total cost?

A   £740
B   £666
C   £660
D   £594

4   The family plan to take the equivalent of £500 in Euros spending money for their four-day break. When ordered in advance, their bank does not charge commission fees on foreign currency exchange.

Exchange rate: 1 Euro is equivalent to £0.79

How many Euros and cents does the family receive in exchange for their £500?

A   €395.00
B   €507.90
C   €632.91
D   €790.00

5   On one night during the Ski Club four-day break in Chamonix, the outside temperature is –18°C. The temperature inside is 27°C.

What is the temperature difference between the inside and the outside?

A   45°C
B   36°C
C   27°C
D   9°C

6 From October to December a total of 2098 people visited Silkstone Ski Sport Centre. Of the total, 681 visitors were under 18 years of age. The rest were 18 years old or over.

What is the approximate ratio of the under 18s to the over 18s?

A 2 : 1
B 1 : 2
C 7 : 3
D 3 : 7

Questions 7 to 9 are about analysis of a paralympics road-race training times.

A group of 30 junior wheel chair racing athletes take part in a training race over 1km at the start of their competition season. The table shows the racers' finishing times.

Seconds	130 – 134.9	135 – 139.9	140 –144.9	145 – 149.9	150 – 154.9
No of Racers	3	6	12	7	2

7 How many racers finished in less than 145 seconds?

A 28
B 22
C 21
D 9

8 In the race 40% of the racers finished in:

A more than 140 seconds
B less than 135 seconds
C between 150 – 154.9 seconds
D between 140 – 144.9 seconds

9 What fraction of the racers finished in between 135 to 144.9 seconds?

A 1/5
B 2/5
C 3/5
D 4/5

Please go on to the next page

Questions 10 to 12 are about the packing of soft drink cans for despatch.

The diagrams show a top view of a packing case and one can.

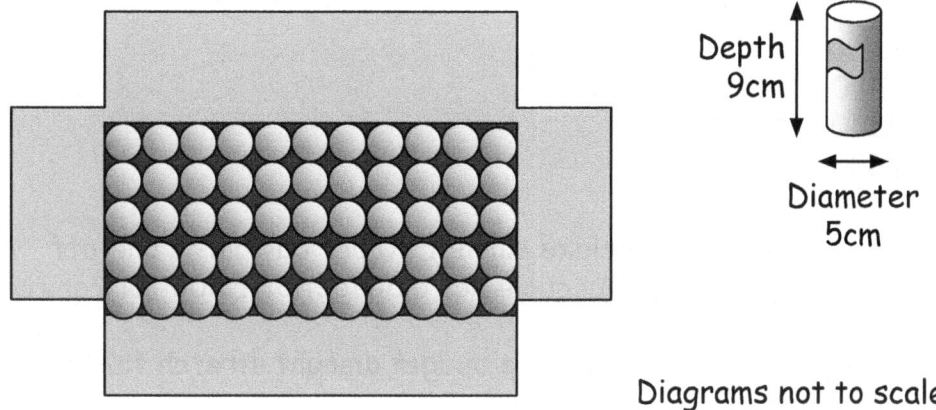

Depth
9cm

Diameter
5cm

Diagrams not to scale

10   There are 3 layers of cans in the packing case. A corner shop orders 4
     cases of the soft drink. How many cans in total will be delivered?

     A   660
     B   495
     C   330
     D   165

11   Using the can measurements, the number of cans in a layer and the number
     of layers of cans, the dimensions of the packing case can be calculated.
     Which of the following are the correct dimensions of the packing case?

     A   50cm x 20cm x 25cm
     B   55cm x 25cm x 27cm
     C   11cm x 5cm x 3cm
     D   55cm x 27cm x 3cm

12   To find the volume of one can, use the formula:

     Volume = Area of cross-section x Depth

     The area of the can in cross-section is approximately 18.75cm²

     To change square centimetres to millilitres, use the conversion:

     1cm³ = 1ml

     What is the approximate volume of soft drink in one can, rounded up to the
     nearest ml?

     A   158ml
     B   169ml
     C   179ml
     D   188ml

Questions 13 to 16 are about car rental on a budget.

13 Rental charge for a small car is £15 per day, plus a one-off preparation fee of £25. What is the cost to rent a car for seven days?

    A   £280
    B   £190
    C   £140
    D   £130

14 The rental charge for a deluxe car is £20 per day plus a one-off preparation fee of £40. A couple has a budget of £300 for car rental.

How many days car rental will the budget amount stretch to?

    A   5 days
    B   11 days
    C   13 days
    D   15 days

15 The car rental company uses a formula to work out the cost per day (D) times the number of days (N) plus the preparation fee (P), in pounds.

Which is the correct formula?

    A   £ = (P × N) + D
    B   £ = (D × N) + P
    C   £ = (P + D) + N
    D   £ = (D + P) × N

16 A family travels 600 kilometres in total during their car rental period and they use 48 litres of fuel costing £1.16 per litre.

What is the cost, to the nearest penny, of the fuel used to travel 1 kilometre?

    A   9p/km
    B   11p/km
    C   12.5p/km
    D   14.5p/km

Please go on to the next page

Questions 17 to 20 are about patient health monitoring in a nursing home.

17 The graph shows the monitored depth of water in a bath in a nursing unit. With the aid of a hoist the patient is lowered into the bath. This takes about one minute.

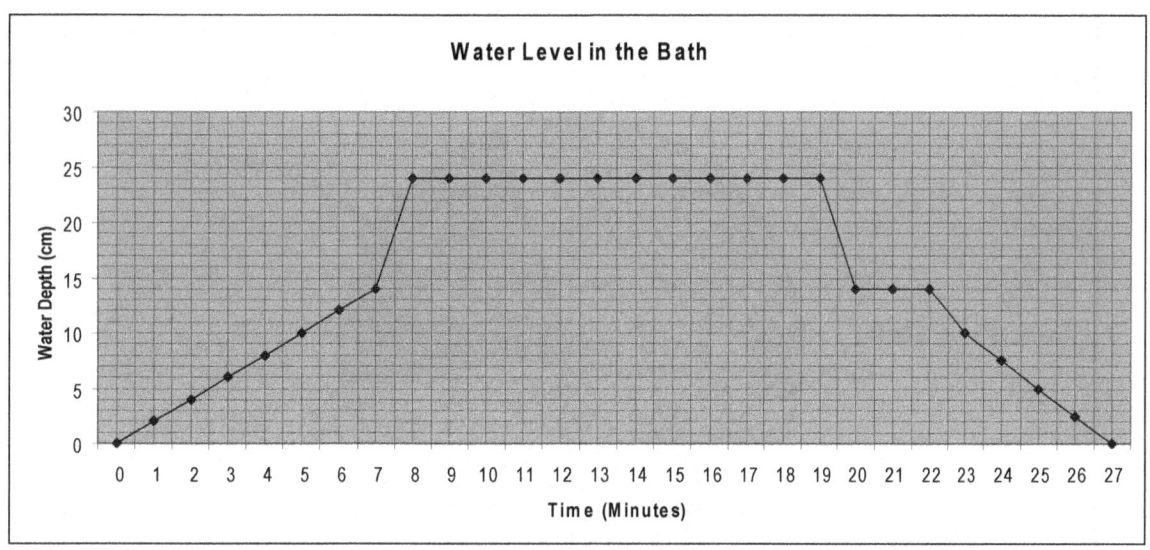

Looking at the graph, which statement is correct?

A   the patient was in the bath for 24 minutes
B   the patient was in the bath for 10 minutes
C   the bath takes more time to fill than it does to empty
D   the bath takes less time to fill than it does to empty

18 A nursing assistant asks a female patient what her weight is in kilograms but she only knows that her weight is about 8 stones and 12 pounds.

> **1 stone = 14 pounds.   1 kilogram = 2.2 pounds (approximately)**

What is the patient's approximate weight in kilograms, to the nearest kilogram?

A   56kg
B   57kg
C   58kg
D   60kg

Please go on to the next page

19   A patient with limited mobility is advised to increase his intake of fibre in his diet. The chart shows the proportions of nutrients in a bowl of breakfast cereal:

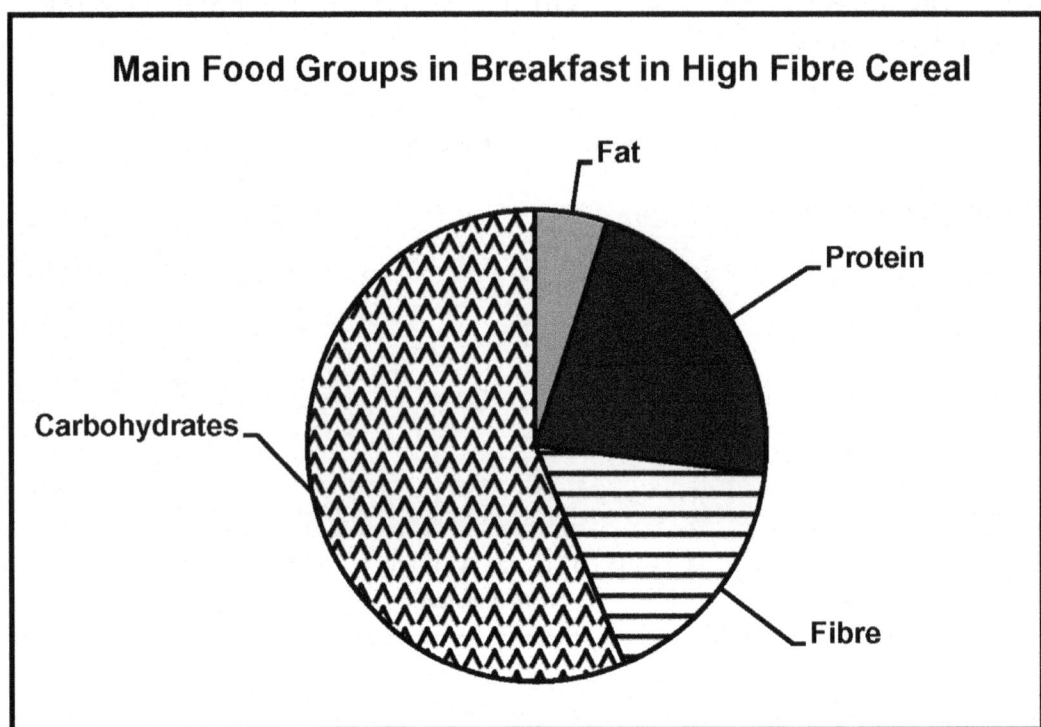

Main Food Groups in Breakfast in High Fibre Cereal

Which of the following descriptions gives the correct gram amount of each food group in the chart?

A   Carbohydrates 3.5g, Protein 10g, Fat 13g, Fibre 33.5g
B   Carbohydrates 33.5g, Protein 3.5g, Fat 10g, Fibre 13g
C   Carbohydrates 13g, Protein 33.5g, Fat 10g, Fibre 3.5g
D   Carbohydrates 33.5g, Protein 13g, Fat 3.5g, Fibre 10g

20   After a long illness a patient is advised to increase her body weight from $7\frac{1}{2}$ stones to 9 stones.

What fraction of her current body weight must she gain to reach her ideal weight?

A $\frac{1}{10}$

B $\frac{1}{5}$

C $\frac{1}{4}$

D $\frac{1}{3}$

Please go on to the next page

Questions 21 and 22 are about updating the Mayor's suite at the City Centre.

21   The diagram shows the area of the ground floor entrance hall to the Mayor's suite that needs new carpet.

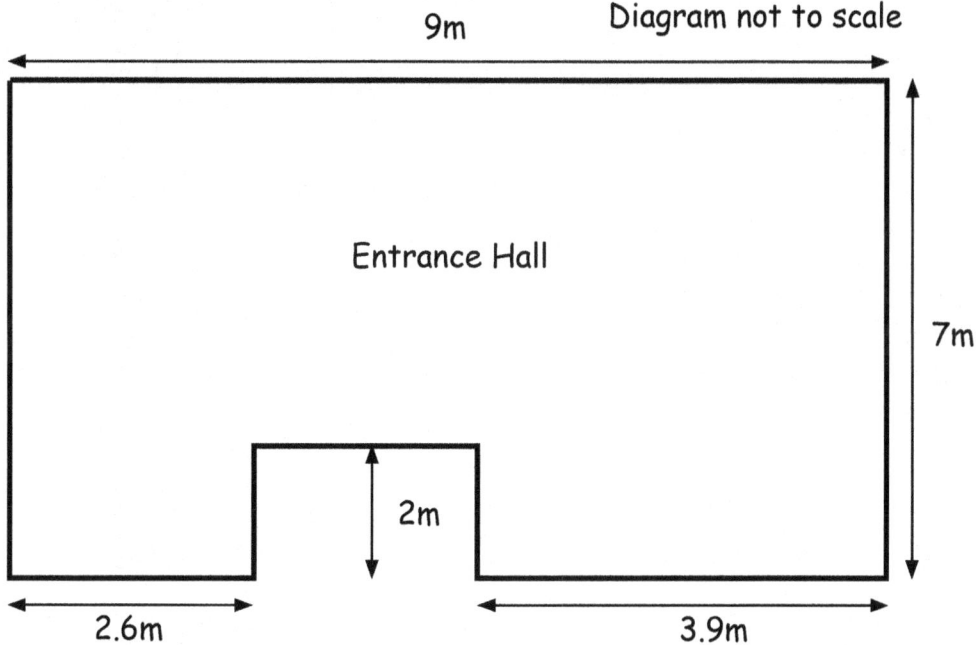

What is the total area of carpet in the entrance hall?

A   63m²
B   61m²
C   58m²
D   53m²

22   The civic emblem on a plaque in the entrance hall requires a new circle of gold leaf to restore its original condition. The circle is 9cm in diameter.

> The area of a circle = π × r²
> where π (pi) is 3.14 and r is the radius

What is the area of the circle of gold leaf to restore the emblem, rounded to two decimal places?

A   20.25cm²
B   63.59cm²
C   81.00cm²
D   254.34cm²

Please go on to the next page

Questions 23 to 27 are about Steelworx sculpture design studio.

The specification for a stainless steel sculpture commissioned by The Electro Theatre for the theatre courtyard is shown in the scale drawing.

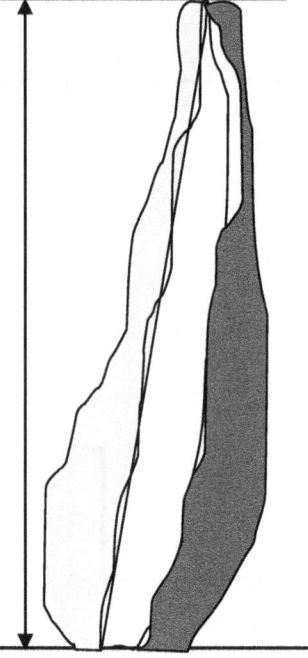

Scale 1 : 50

Full-size Weight = 250kg

Stainless Steel Composition
Iron   74%
Carbon   14%
Chromium   12%

23 What is the actual height of the sculpture in metres?

    A   4.3m
    B   4.8m
    C   8.3m
    D   8.6m

24 One metric tonne of the stainless steel costs £4290. (1 tonne = 1000kg)

What is the cost of the stainless steel for this sculpture?

    A   £429.50
    B   £859.00
    C   £1072.50
    D   £1429.00

25 The full-size sculpture weighs 250kg. 74% of this weight is Iron. What is the proportionate weight of Iron in the Stainless Steel?

    A   175kg
    B   185kg
    C   195kg
    D   205kg

26 Steelworx deliver and install sculptures on-site. The table shows the costs.

Delivery	Installation
Free up to 50km from studio; £1 per km over 50km	£100 + VAT up to 4 hours
	£30 + VAT per hour over 4 hours (or part of)
	VAT @ 17.5%

The Electro Theatre site is 45 kilometres from the studio. Installation of the sculpture on-site takes 7 hours and 45 minutes. Which calculation shows the total cost of delivery and installation, in pounds?

A (100 + 30 x 3) x 1.175
B (100 + 30 x 4) x 1.175
C 100 + (30 x 3) x 1.175
D 100 + (30 x 4) x 1.175

27 The Electro Theatre site occupies an area of 1/5 of one hectare.

One hectare = (100 x 100)m²

The Electro Theatre site

width = 40m

Diagram not to scale

What is the length of the Electro Theatre site?

A 20m
B 40m
C 50m
D 80m

Questions 28 and 29 are about a working trip to Japan.

28 After his university studies, Andrew travels to Japan to teach English. He receives a bursary of 24,500 Japanese Yen (¥) per week. He has a contract to teach for 13 weeks.

£1 = ¥196

How much money, in pounds, does Andrew receive for the 13-week period?

A   £3694
B   £2450
C   £1684
D   £1625

29 After working 13 weeks, Andrew plans a trip by boat from Aomori to Hakodate. The line on the scale drawing shows the distance between the ports on the islands of Honshu and Hakkaido.

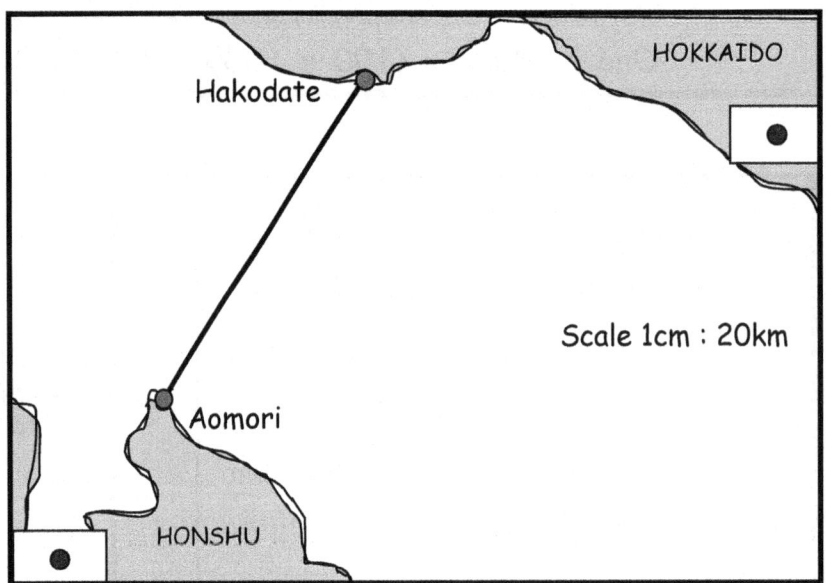

What is the actual distance between Aomori and Hakodate?

A   96km
B   87km
C   78km
D   59km

Please go on to the next page

Questions 30 to 34 are about a horticultural development laboratory.

A new variety of fast growing soya bean is being developed for commercial production. The heights of 20 two week-old seedlings are recorded in the table in cm.

3.4	3.8	3.2	3.8	3.4	3.7	4.5	2.7	3.7	3.5
4.1	3.4	3.5	3.1	2.7	3.9	3.4	4.6	3.4	4.2

30 What is the mean of this sample of seedlings?

- **A** 3.1cm
- **B** 3.4cm
- **C** 3.6cm
- **D** 4.1cm

31 What is the median of this sample of seedlings?

- **A** 3.2cm
- **B** 3.3cm
- **C** 3.4cm
- **D** 3.5cm

32 What is the modal seedling height in this sample?

- **A** 3.4cm
- **B** 3.5cm
- **C** 3.6cm
- **D** 3.8cm

33 What is the range of heights recorded?

- **A** 4.6cm
- **B** 3.7cm
- **C** 2.9cm
- **D** 1.9cm

34 Out of a total of 500 seeds planted, 485 germinated successfully. What is the ratio of non-germinated seeds to germinated seeds?

- **A** 15 : 500
- **B** 9 : 194
- **C** 3 : 97
- **D** 1 : 97

Please go on to the next page

Questions 35 and 36 are about second-hand car dealer's sales in two years.

The chart shows the number of second-hand cars sold in 2006 and 2007.

In 2006, a total of number 90 Rover and Vauxhall cars were sold.

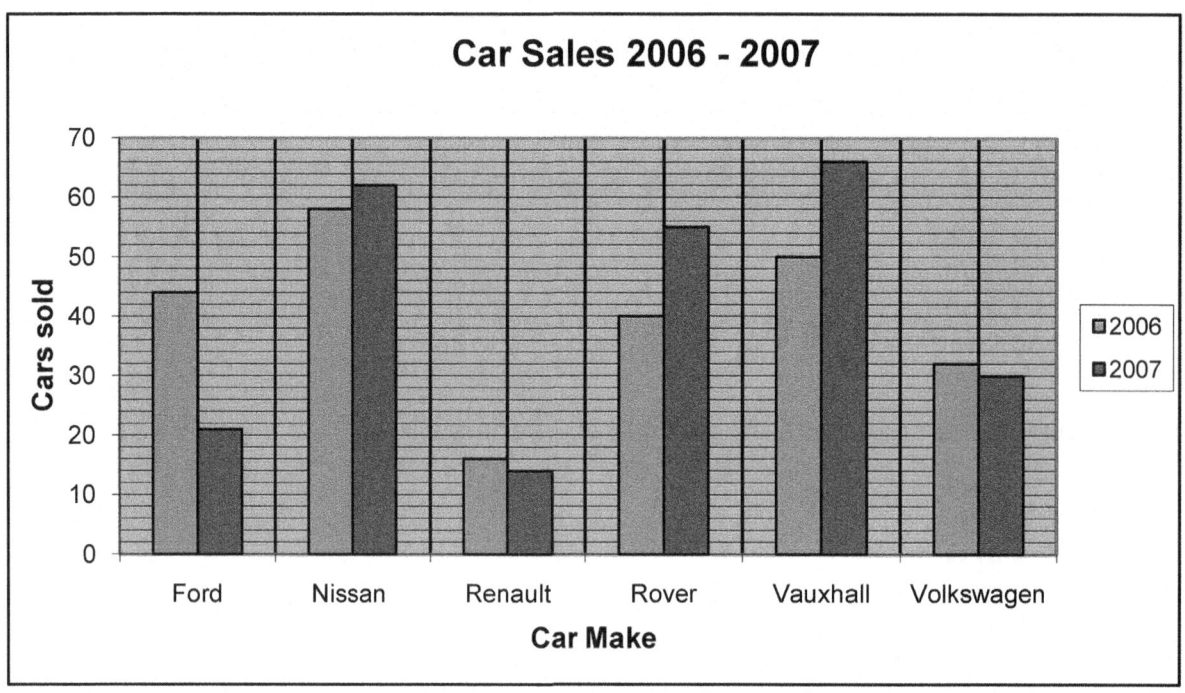

35   How many more Rover **and** Vauxhall cars were sold in 2007?

   A   33
   B   31
   C   16
   D   15

36   Referring to the chart, which one of the statements is true?

   A   sales of all makes of car decreased from 2006 to 2007
   B   sales of all makes of car increased from 2006 to 2007
   C   sales of Ford, Renault and Volkswagen decreased from 2006 to 2007
   D   sales of Ford, Renault and Volkswagen increased from 2006 to 2007

Please go on to the next page

Questions 37 and 38 are about activities at an international sports event

37 A sport standards official measures the perimeter of the running track, as shown in the diagram.

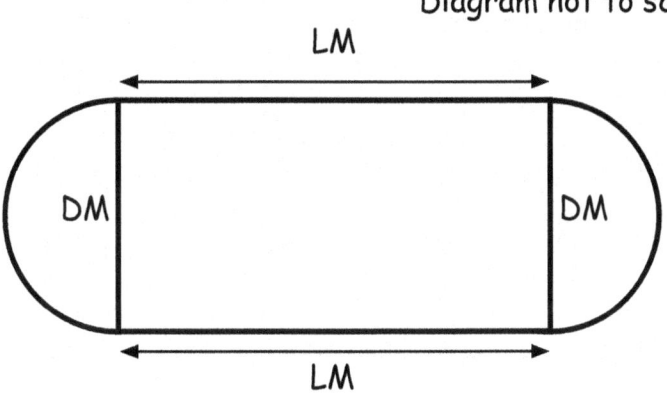

Dm = diameter (metres)
Lm = length (metres)

> Perimeter of a circle = π × diameter

Which calculation finds the perimeter of the running track, in metres?

A (2 x Lm) x (π x Dm)
B (2 x Lm) + (π x Dm)
C (2 x π) x (Dm x Lm)
D (2 x Dm) + (Lm x π)

38 At a sports event, 1400 tickets were available on the day.

72.5% of the tickets were sold. Tickets cost £5 each.

How much money, in pounds, was received from the sale of 72.5% of the tickets?

A £5075
B £5600
C £5880
D £5985

Please go on to the next page

Questions 39 and 40 are about vehicles counted at a busy road junction.

39 The chart shows the results of a traffic monitoring survey:

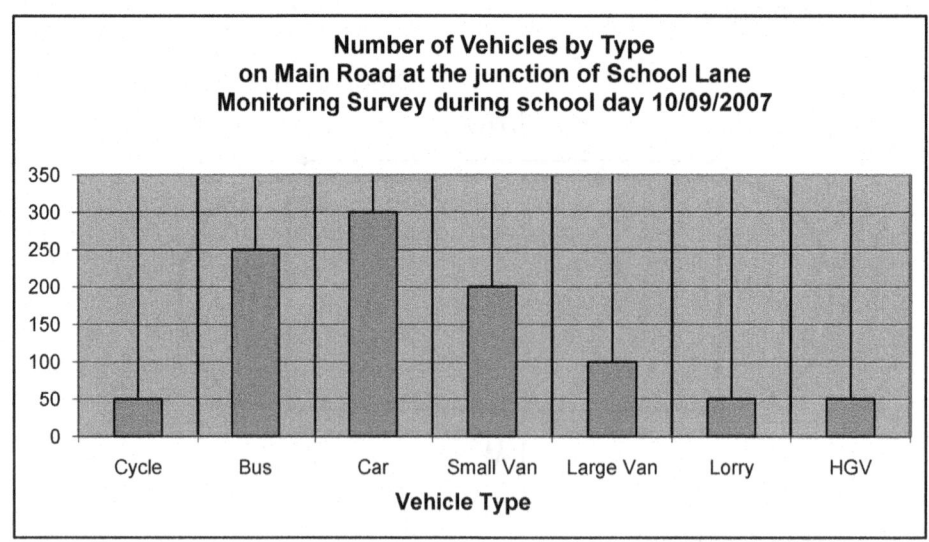

What number of goods vehicles (vans, lorries and heavy goods vehicles) was recorded in the survey?

A 600
B 400
C 200
D 50

40 During the school day 1000 vehicles use the junction at School Lane.

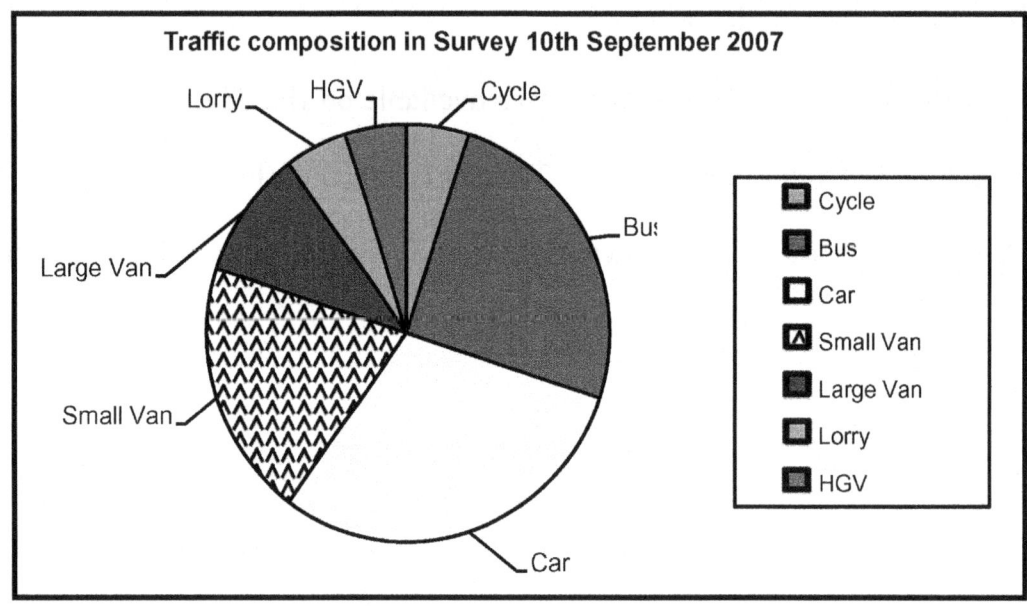

How many degrees, out of the 360° of the pie chart circle, are used to represent the proportion of buses recorded in the survey?

A 36°
B 72°
C 90°
D 108°

End of Paper

Practice Multiple-choice Paper
Suitable for:

Key Skills Level 2 Application of Number
Level 2 Adult Numeracy

Paper Twelve

YOU NEED

- This test paper.
- A pen.
- A pencil and eraser.
- An Answer Sheet.
- A ruler marked in centimetres and millimetres

You may NOT use a calculator.
You may use a bilingual dictionary.
There are 40 questions on this paper. Try to answer ALL the questions.
When you have completed the questions you must check your answers, then check them again.

YOU HAVE QUARTER OF AN HOUR TO READ THE PAPER
AND ONE HOUR TO COMPLETE THE 40 QUESTIONS

INSTRUCTIONS

- Make sure you write your name and today's date on the Answer Sheet. Use a pen to do this.
- Use a pencil to mark your answers so if you change your mind you can erase your choice and select another.
- Make sure that for each question you have only selected one answer. If you select more than one, the answer will not be marked.
- Read each question carefully before you select an answer.

Note for learners and tutors: This is a practice test that has been designed to closely resemble the questions and question styles of a "live" paper.

Questions 1 to 4 are about a survey conducted by Fit-4-Life Sports and Fitness Club.

The table shows numbers of men and women members visiting the club during one week.

Session Time	Fit4Life Sports and Fitness Club						
Men	Mon	Tue	Wed	Thu	Fri	Sat	Sun
8am - 12 noon	57	49	65	59	73	50	42
12 noon - 4pm	65	62	48	79	52	61	57
4pm - 8pm	60	53	59	42	46	33	31
Women	Mon	Tue	Wed	Thu	Fri	Sat	Sun
8am - 12 noon	36	30	29	22	36	48	31
12 noon - 4pm	50	55	58	30	66	39	37
4pm - 8pm	87	82	73	20	49	88	71

1   Referring to the table, what is the ratio of men to women visiting the club during all sessions on Thursday?

A   3 : 1
B   5 : 2
C   9 : 5
D   18 : 7

2   What is the range of the numbers of both men and women visiting the club on Friday?

A   27
B   30
C   37
D   46

3   What is the median number of the women members who visited the club between 12 noon and 4pm from Monday to Sunday?

A   50
B   39
C   37
D   30

4   On which day and session time did the lowest number of men visit the club?

A   Mon   4pm - 8pm
B   Thu   8am - 12 noon
C   Sat   12 noon - 4pm
D   Sun   4pm - 8pm

Questions 5 to 7 are about Fit-4-Life Sports and Fitness Club membership.

5   The annual subscription to Fit-4-Life Sports and Fitness Club is currently
     £235. The club is offering a special discount of 15% on annual membership
     to anyone signing up in the next 14 days.

     What is the discounted cost of annual membership?

     A   £220.00
     B   £215.00
     C   £199.75
     D   £115.15

6   Gina, member No. 2320, is working out on the treadmill machine, which
     is set to end at 10:45. She looks at the timer panel, which displays the
     following message:

| Member No | 2320 | You have | 27 | minutes remaining |

     What time is it when Gina looks at the timer?

     A   11:47
     B   11:18
     C   10:53
     D   10:18

7   One circuit of the club's indoor running track measures 50m. Club member
     Paula runs three-and-a-half circuits in 5 minutes.

     > 1 metre = approximately 39 inches
     > 1 foot = 12 inches

     Approximately, what is the distance that Paula runs in 5 minutes, in feet
     rounded to the nearest foot?

     A   550 feet
     B   569 feet
     C   578 feet
     D   583 feet

Please go on to the next page

Questions 8 to 14 are about production trials of new plant varieties at a horticultural research station.

The diagram shows two seed trays in a production trial and the numbers of seeds that germinated, (X), and seeds that did not germinate, (O) in the last seven days.

Tray A	Tray B
X O X O O X X X X X X X	O X X O X O X X X O X X
O X O X X X X X X X X O	X O X X O O X O X X O O
X X X X X X X X X O X O	X O X X O X X X X O O X
O X X X X X X X X X O X	O X O X X X O X X X O X
O X O X X X X X X X X O	X X X O X X X X X O X X
X O X X O O X O X X O X	X O X O X O O X O O O X
X X X O X X X X X O O X	X X O X X O X X X O O X
X X X X X X O X O X O X	O X O X X X O O X X X O

8 What fraction of the seeds that were planted in Tray A failed to germinate?

    A   1/6
    B   1/5
    C   1/4
    D   1/3

9 What percentage of the seeds in tray B germinated successfully?

    A   65.0%
    B   62.5%
    C   37.5%
    D   35.0%

The table shows the heights, in metres, of 10 plants in a production trial of a variety of vine tomatoes 28 days after germination.

1.16	1.03	1.18	1.11	1.18
1.18	1.17	1.19	1.14	1.06

10 What is the mean height of the vine tomato plants?

    A   1.14m
    B   1.15m
    C   1.16m
    D   1.17m

11 What is the modal height of the tomato plants

    A   1.19m
    B   1.18m
    C   1.06m
    D   1.03m

12 In the trial, the tomato plants were measured to the nearest centimetre. What is the greatest possible height of a tomato plant measured as 1.16m?

A 1.164m
B 1.165m
C 1.169m
D 1.174m

13 Soil plots used in tomato production trials measure 5m x 5m. Seaweed-based fertiliser is applied to soil plots at a rate of 80g per square metre. What number of soil plots can be fertilised from one 25kg bag of fertiliser?

A 8.5
B 10
C 12.5
D 15

14 Each soil plot is stocked with 22 of the vine tomato plants to grow-on to full-size. Each plant is expected to yield at least 6 pounds of tomatoes.

> 1kg = 2.2 pounds

What is the minimum yield weight of tomatoes, in kilograms, from one soil plot?

A 32.2kg
B 60.0kg
C 79.2kg
D 132.0kg

Questions 15 to 22 are about a restaurant business in the Channel Island of Jersey.

15 A restaurant director takes over a restaurant in Gorey on the island of Jersey. The table shows the financial status of the restaurant during the last 3 months:

Restaurant Mange Frais, Rue Sancerre, Gorey, Jersey			
Balance Sheet	April	May	June
Starting Balance	£1,885	£1,989	-£1,003
Profit / Loss	£104	-£2,992	-£168
End Balance	£1,989	-£1,003	

What was the End Balance for June?

A -£1,171
B £1,171
C -£835
D £835

16 The restaurant director travels to mainland France by boat to meet with new suppliers of wine and cheese. The line on the scale drawing indicates the distance between the port in Jersey and the destination, Carteret.

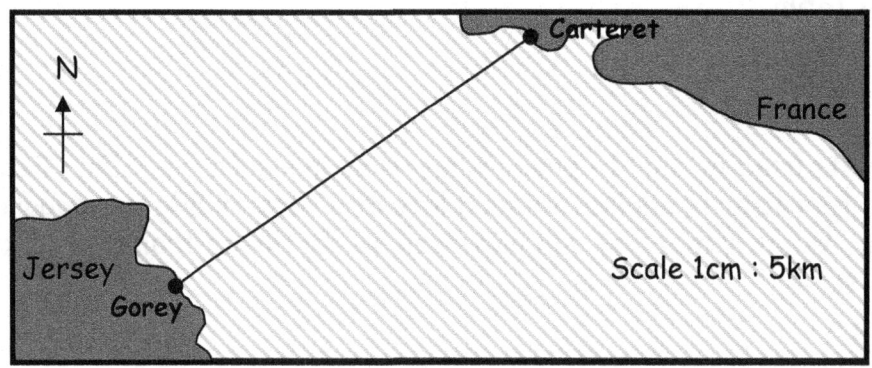

What is the distance between Gorey and Carteret?

A   26.5km
B   28.5km
C   52.0km
D   54.0km

17 The restaurant director buys a quantity of a type of regional cheese in the shape of a flat cylinder with dimensions as shown in the diagram.

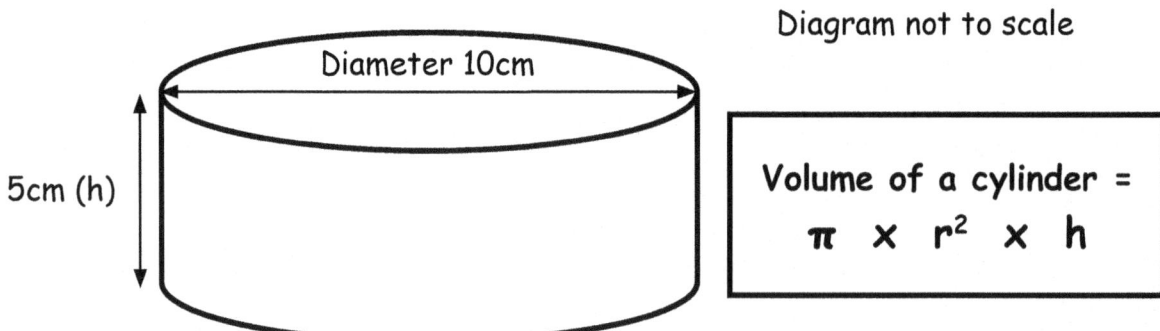

Diagram not to scale

Volume of a cylinder =
$\pi \times r^2 \times h$

Which of the following calculations finds the volume, in square centimetres, of the cylinder shaped cheese?

A $\pi \times 10 \times 5$
B $\pi \times 10 \times 10 \times 5$
C $\pi \times 5 \times 5 \times 5$
D $\pi \times 5 \times 5 \times 10$

18 Another type of regional cheese bought by the restaurant director has a similar shape. An individual cheese has the volume of 191cm³.

1cm³ of this cheese weighs 2.38g. What is the total weight of the cheese, rounded to the nearest gram?

A   455g
B   454g
C   453.6g
D   453.63g

19 A red house wine that the restaurant director selects is sold in barrels holding 22.5 litres. The wine is to be served to his customers in 750ml jugs. How many 750ml jugs can be filled from one barrel of the red wine?

A   22
B   22.5
C   30
D   32.5

20 At trade price, one 22.5-litre barrel of red costs 170 Euros.

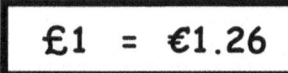

£1 = €1.26

What is the cost of one barrel of the red wine in British pounds and pence?

A   £126.00
B   £134.92
C   £214.20
D   £225.00

21 The restaurant director buys a quantity of white wine that costs £3.25 per 70cl bottle. The restaurant's wineglasses contain 140ml and one glass of white wine sells for £2.45.

100cl = 1 litre

How much profit does the restaurant make from each bottle of the white wine?

A   £16.25
B   £14.00
C   £12.25
D   £9.00

Please go on to the next page

22   A customer asks for the bill after his meal at Restaurant Mange Frais.
     The part completed bill is shown here:

Table No 7	Food Order	Totals
Quantity		
1	Breaded Mushrooms	£3.75
1	Boeuf Burgingnon	£12.50
1	Dessert Camembert	£5.75
1	House Red, glass	£2.75
1	Coffee	£1.75
	Add Service Charge: 12.5 %	

What is the total cost of the meal, including service charge?

A   £23.50
B   £26.50
C   £29.81
D   £31.81

Questions 23 to 27 are about second and third year college students.

23   In a survey of 720 student and college staff, one half of them travel to
     the college by bus, one quarter travel by city metro, one sixth walk to
     college and the rest arrive by bicycle.

     What fraction of the 720 students and staff arrives at college by bicycle?

A   1/12
B   1/6
C   1/5
D   1/4

24   A class of 72 second year college students sit a maths exam. Five eighths
     of the class pass the exam. What number of students **did not** pass the
     exam?

A   28
B   27
C   25
D   24

25   There are 325 third year catering students at the college. 195 of the
     students are male. What percentage of the students is female?

A   66%
B   55%
C   40%
D   37%

26 There are 325 third year catering students at the college. 195 of the students are male. What fraction of the students are male?

    A   3/5
    B   2/5
    C   1/4
    D   1/3

27 There are 325 third year catering students at the college. 195 of the students are male. What is the ratio of female to male students?

    A   5 : 3
    B   3 : 5
    C   3 : 2
    D   2 : 3

Questions 28 to 32 are about modern domestic bread baking.

28 The landlady of a guesthouse decides to buy a bread-making machine. She retrieves the following list of products on an Internet comparison site.

Make / Model	Colour	Number of Loaf Sizes	Fast Bake (minutes)	Baking Programmes	Price
Antony W Thompson	Grey	6	68	19	£99.49
Cookworks	White	2	58	12	£23.99
Cookworks	St. Steel	3	58	13	£29.99
Morphy Richards FB	White	2	88	10	£39.99
Morphy Richards Compact	White	3	88	12	£44.47
Morphy Richards FB Plus	St. Steel	4	78	13	£69.76
Morphy Richards FB Super	St. Steel	4	68	15	£75.99
Panasonic SD-264	White	3	58	17	£79.99
Panasonic SD-265	White	3	78	19	£99.99

The landlady wants a machine that can fast-bake in less than 1 hour, has three or more bread sizes and is white to match her kitchen colour scheme.

What is the price of the machine that meets all of the landlady's requirements?

    A   £23.99
    B   £75.99
    C   £79.99
    D   £99.99

29   The landlady finds that the instruction manual with her new bread-making machine has metric measures only. Her recipe for a traditional loaf requires one and a half pounds (1lb 8oz) of flour.

1lb = 16 ounces	1oz = 28.4g

What is 1lb 8oz in grams, rounded to the nearest gram?

A   681g
B   682g
C   766.6g
D   767g

30   The bread-making machine has a timer and does a fast bake in 58 minutes. The ingredients are placed into the baking compartment and left there until the programmed baking time begins. The landlady wants the bread to be ready at a quarter to eight the next morning.

What start time should she program into the timer to start baking?

A   6:43am
B   6:47am
C   7:43am
D   7:47am

31   The ingredients to bake a small white loaf in the bread-maker include 500g of strong white flour per loaf. The flour is packaged in a 25-kilogram sack. How many small loaves can be made from one 25kg sack?

A   12.5
B   25
C   50
D   75

32   Organic granary bread flour costs £16 for a 12.5kg sack. The recipe for a speciality bread uses 250g of the granary flour per small loaf. The shop bought loaf, equivalent in size to the speciality bread, costs £1.12.

How much cheaper is a machine baked granary loaf than the shop bought equivalent?

A   80p cheaper
B   70p cheaper
C   32p cheaper
D   12p cheaper

Questions 33 and 34 are about trends in healthy lifestyle.

33 A dietician wants to draw a pie chart to represent the recommended percentage weight of food groups in a healthy diet as shown in the table.

Food Groups	Percentage
Meat, Fish, Eggs and Beans	15
Bread, Rice, Potatoes, Pasta	30
Fruit, Vegetables and Fibre	30
Milk and Dairy Foods	15
Foods High in Fat or Sugar	10

How many degrees will the dietician use to represent Foods High in Fat or Sugar?

A   45°
B   36°
C   30°
D   10°

Please go on to the next page

34 "Smoking in the UK is decreasing but by less each year". Which of the following graph shapes represents this statement most accurately?

Graph No 1

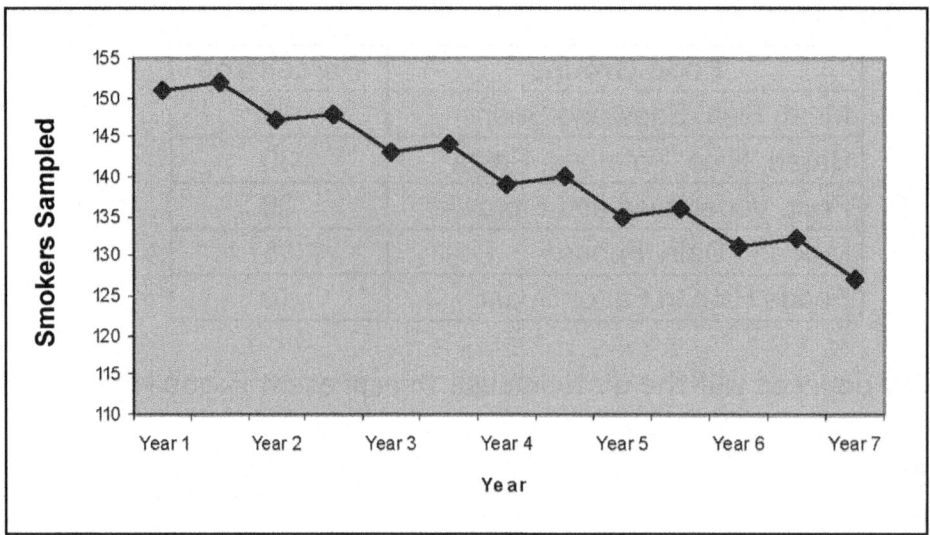

Graph No 2

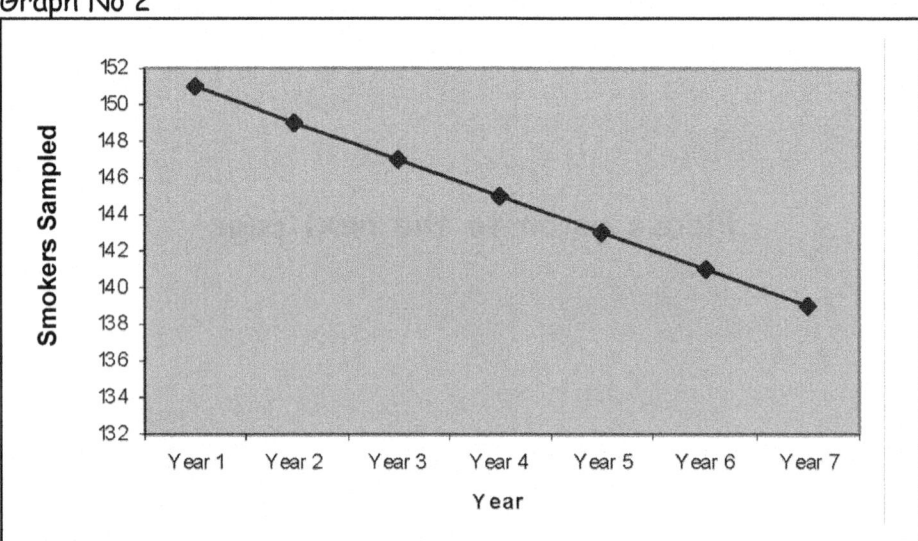

Graph No 3

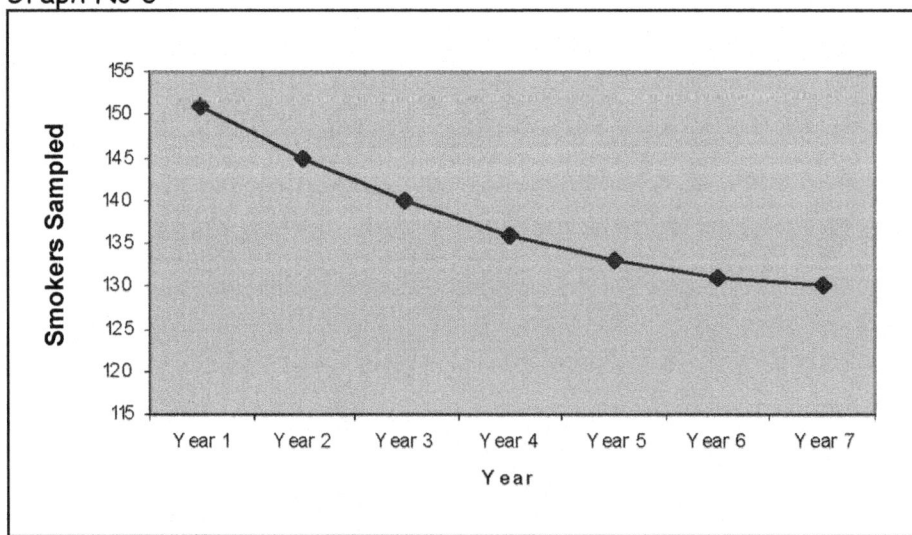

Graph No 4

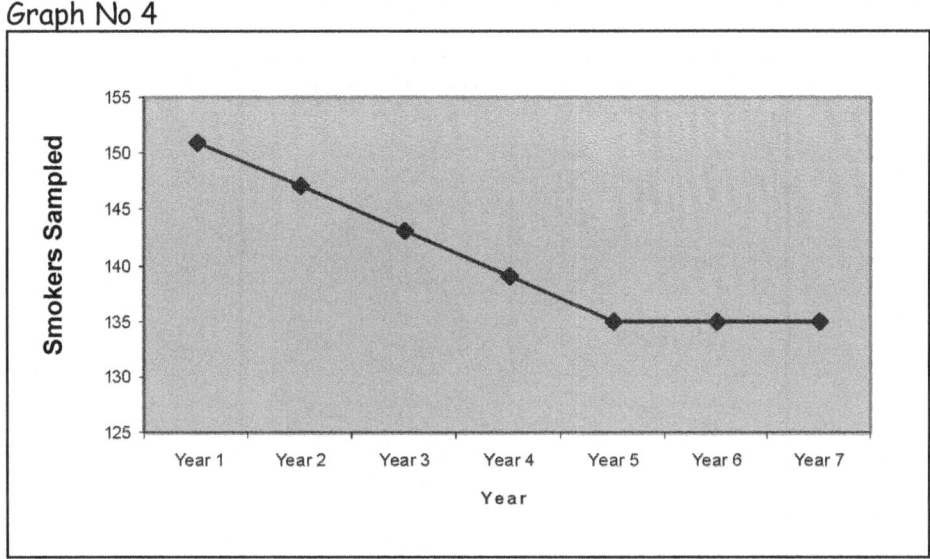

A   Graph No 1
B   Graph No 2
C   Graph No 3
D   Graph No 4

Questions 35 and 36 are about a student's research for a Health and Social Care assignment.

35   Sonja surveys the other students in her class to ask how many siblings (sisters and / or brothers) each has. She draws a bar chart to display the results:

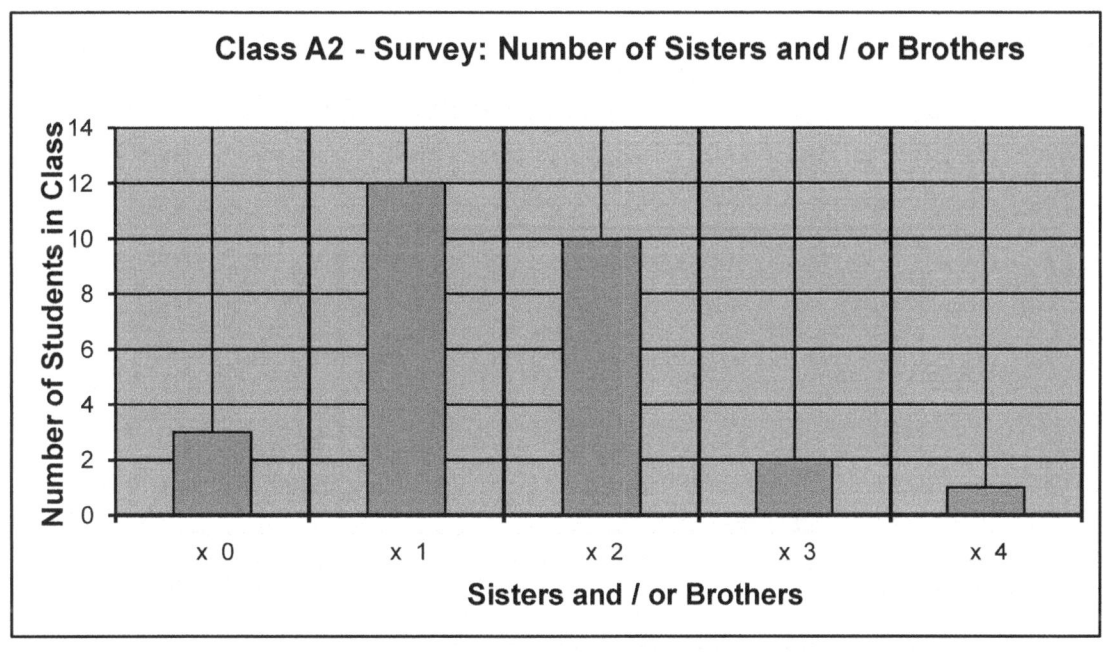

What number of students in her class has two siblings or more?

A   25
B   13
C   12
D   10

36   Sonja surveys ages of her classmates and draws a pie chart of the results.

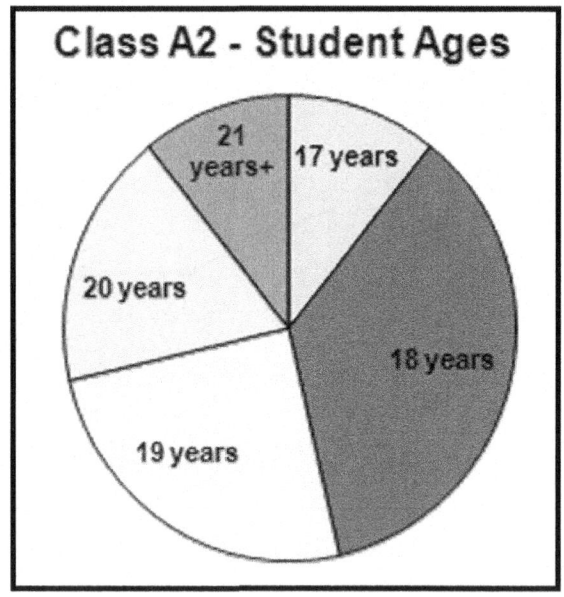

Which one of the following statements is true?

A   the majority of students are over 21 years old
B   the majority of students are under 18 years old
C   the majority of students are 20 years old
D   the majority of students are 18 years old

Questions 37 and 40 are about a suburban garden.

The diagram shows the dimensions of a garden in the suburbs of a city:

Diagram not to scale

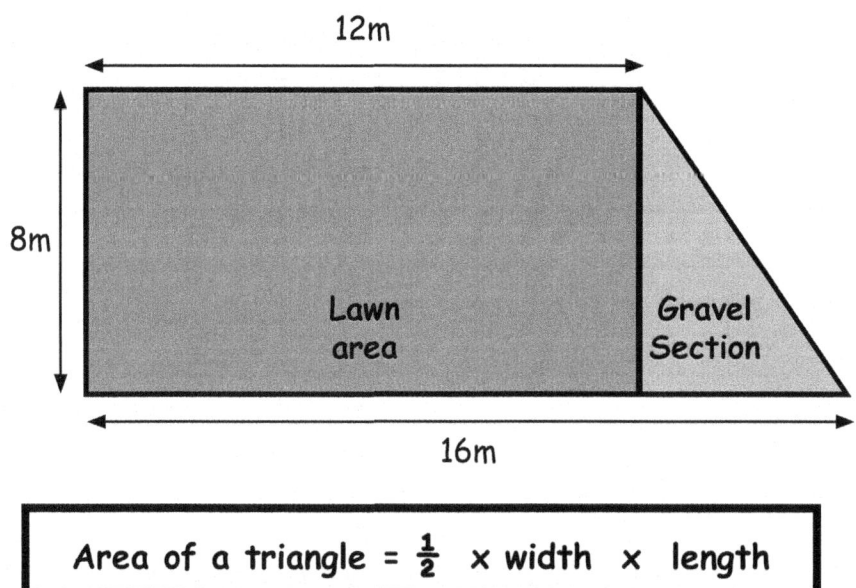

Area of a triangle = ½ × width × length

37   What is the area of the Gravel section?

    A $16m^2$
    B $14m^2$
    C $12m^2$
    D $8m^2$

38   A drainage channel is to be cut around the perimeter of the Lawn area. What is the total length of the perimeter?

    A   36m
    B   38m
    C   40m
    D   44m

39   The table shows minimum temperatures recorded in a garden during a four-week period in March.

Temp °C	Mon	Tue	Wed	Thu	Fri	Sat	Sun
Week 1	0	-1	3	4	2	3	1
Week 2	-1	0	-3	4	2	3	0
Week 3	1	3	4	2	3	2	3
Week 4	3	4	5	3	3	6	3

When the outside temperature drops to 2°C or less, a thermostat switches on a heater.

On what fraction of days was the heater switched on?

    A   2/7
    B   3/7
    C   4/7
    D   5/7

40   What is the temperature range during the four-week period?

    A   9°C
    B   6°C
    C   3°C
    D   -3°C

End of Paper

Key Coverage

Description / Key Coverage: Paper 1	In Question No.	Also in Q No.
Tables / Charts / Graphs	15, 16, 18, 29, 30, 39, 40	4, 5, 14, 24, 29
Fractions / Decimals / Percentages	13, 14, 38	9, 37
Amounts / Proportion	16, 21, 30, 31, 34	14, 15, 17, 30
Mean / Mode	11	
Median / Range	4, 12, 27	
Metric / Imperial Measures / Conversion	7, 8, 35, 36	22, 25, 29
Currency	9, 18, 20	12, 13, 23, 38
Time / Temperature	1, 2, 3, 4, 5, 40	13, 15, 25, 36
Weight / Capacity	27, 28, 29, 32, 33	34
Area / Perimeter / Volume	23, 24, 25	37
Ratio / Scale	6, 10, 22, 28	
Levels of Accuracy / Estimation	19, 20, 26, 32, 38	5, 9, 21, 30, 35
Formulae	17, 36, 37	13

Description / Key Coverage: Paper 2	In Question No.	Also in Q No.
Tables / Charts / Graphs	1, 4, 8, 15, 28, 31, 39, 40	26
Fractions / Decimals / Percentages	1, 3, 6, 7, 10, 21, 28, 30, 35	11, 12, 22, 25, 31
Amounts / Proportion	4, 8, 13, 14, 32	7, 19, 34, 35, 37, 40
Mean / Mode	5, 18	
Median / Range	2, 4	
Metric / Imperial Measures / Conversion	19, 23, 37	21, 14, 29, 34
Currency	8, 9	10, 13, 15
Time / Temperature	2, 16, 38	1, 3, 28
Weight / Capacity	27, 29, 34, 36	30
Area / Perimeter / Volume	17, 24, 25, 26, 33, 34	36
Ratio / Scale	11, 12, 16, 22	
Levels of Accuracy / Estimation	3, 20, 23, 27, 29, 33, 40	10, 13, 15, 24
Formulae	17, 24, 26	

Description / Key Coverage: Paper 3	In Question No.	Also in Q No.
Tables / Charts / Graphs	4, 8, 18, 19, 27, 28, 29, 31, 34	17, 32, 33, 36, 38, 39
Fractions / Decimals / Percentages	2, 11, 19, 20, 37, 39	4, 24, 35
Amounts / Proportion	6, 7, 26	22, 23
Mean / Mode	4	
Median / Range	28, 33	
Metric / Imperial Measures / Conversion	9, 14, 23, 34, 35, 38	3, 7, 10 15, 16, 24, 25
Currency	17, 25, 26, 30, 40	4, 6, 7, 8, 11, 17, 21
Time / Temperature	5, 12, 31, 32, 33, 36	13, 16, 18, 27, 28, 29
Weight / Capacity	7, 34, 38	25
Area / Perimeter / Volume	3, 21, 25, 35	
Ratio / Scale	1, 15, 22, 24	23
Levels of Accuracy / Estimation	2, 10, 19, 20, 26, 31, 38, 40	6, 9, 13, 17, 23, 37, 39
Formulae	12, 13, 16	

Description / Key Coverage: Paper 4	In Question No.	Also in Q No.
Tables / Charts / Graphs	1, 6, 9, 15, 21, 24, 32, 40	7, 8, 33, 38
Fractions / Decimals / Percentages	15, 18, 25	3, 19, 24, 31, 34, 38
Amounts / Proportion	10, 20	9
Mean / Mode	8, 9, 19, 32	
Median / Range	16, 17, 26	
Metric / Imperial Measures / Conversion	3, 4, 5, 28, 35, 36	11, 12, 13, 17, 18, 27
Currency	2, 5, 6, 8, 19, 24, 25, 26, 38	10, 20, 21, 23, 30, 37
Time / Temperature	9, 14, 23, 30, 32, 33, 37	20, 31
Weight / Capacity	4, 11, 13	
Area / Perimeter / Volume	12, 27, 29	
Ratio / Scale	27, 28, 34, 36, 39	37
Levels of Accuracy / Estimation	7, 22, 23, 31	4, 8, 11, 15
Formulae	12, 29	

Description / Key Coverage: Paper 5	In Question No.	Also in Q No.
Tables / Charts / Graphs	2, 11, 14, 21, 22, 23, 36, 40	10, 22, 23, 30, 33, 35
Fractions / Decimals / Percentages	3, 20, 34	11, 18, 21, 22, 25
Amounts / Proportion	7, 8, 9, 12, 13, 25, 31	
Mean / Mode	4, 30	
Median / Range	10, 33	
Metric / Imperial Measures / Conversion	6, 32	7, 17, 19, 39
Currency	3, 15, 18, 22, 25, 31	5, 8, 10, 12, 13, 16, 22
Time / Temperature	14, 15, 24	23
Weight / Capacity	6, 26, 37, 38, 39	28
Area / Perimeter / Volume	28, 29	6
Ratio / Scale	1, 17, 19, 27	
Levels of Accuracy / Estimation	16, 19, 35	1, 21, 24, 29, 34, 39
Formulae	5, 29	

Description / Key Coverage: Paper 6	In Question No.	Also in Q No.
Tables / Charts / Graphs	1, 3, 12, 19, 22, 23, 24, 28	25, 26, 27, 29, 32
Fractions / Decimals / Percentages	18, 21, 23, 26	12, 32, 34
Amounts / Proportion	7, 14, 27, 29, 37	9, 10, 24
Mean / Mode	12	2
Median / Range	13, 19, 34	
Metric / Imperial Measures / Conversion	4, 30, 31, 35, 36, 37, 39	2, 3, 5, 7, 8, 15, 16, 20
Currency	2, 3, 8, 23, 39	9, 10, 27, 29
Time / Temperature	6, 12, 14, 25, 32, 33, 34, 40	1, 13
Weight / Capacity	16, 30	15, 38
Area / Perimeter / Volume	8, 10, 20, 38	
Ratio / Scale	5, 11, 15, 17, 35, 36	
Levels of Accuracy / Estimation	9, 16, 22, 28, 39	6, 17, 21, 24, 31, 37
Formulae	6, 7, 20, 40	

Description / Key Coverage: Paper 7	In Question No.	Also in Q No.
Tables / Charts / Graphs	1, 4, 6, 13, 19, 21, 22, 26, 36	8, 9, 11, 15, 25, 30
Fractions / Decimals / Percentages	1, 6, 14, 23, 37, 38	7, 18, 19, 34
Amounts / Proportion	10, 15, 21, 22	34
Mean / Mode	3, 5, 11, 20	
Median / Range	2, 8, 12, 35	
Metric / Imperial Measures / Conversion	9, 35, 39, 40	10, 14, 16, 28, 31
Currency	19, 24, 29, 30	9, 19, 20, 23
Time / Temperature	25, 26, 27	1, 2, 3, 8, 18
Weight / Capacity	7, 10, 17, 32, 34, 40	
Area / Perimeter / Volume	16, 28, 31	
Ratio / Scale	13, 28, 33, 39	
Levels of Accuracy / Estimation	18, 19, 21, 22, 28, 33, 39	31, 32, 36, 46
Formulae	18, 27, 31	26

Description / Key Coverage: Paper 8	In Question No.	Also in Q No.
Tables / Charts / Graphs	1, 5, 6, 17, 20, 24, 26, 34, 36	11, 13, 23, 25, 27, 30
Fractions / Decimals / Percentages	3, 10, 30, 31, 35, 38	14, 18, 25, 29
Amounts / Proportion	1, 11, 12, 13, 21, 22, 25, 27	17, 18, 33, 36, 37
Mean / Mode	2, 3	
Median / Range	2, 7	
Metric / Imperial Measures / Conversion	14, 29	9, 18
Currency	4, 10, 12, 15, 16, 19, 21, 29, 33	3, 8, 13, 19, 27, 32
Time / Temperature	8, 18, 37	22, 30, 36
Weight / Capacity	9, 14, 24, 40	
Area / Perimeter / Volume	28, 39	40
Ratio / Scale	32	
Levels of Accuracy / Estimation	4, 7, 18, 19, 23, 27	9, 10, 14, 29, 33
Formulae	28, 39	

Description / Key Coverage: Paper 9	In Question No.	Also in Q No.
Tables / Charts / Graphs	4, 5, 6, 21, 28, 31, 33, 39	30, 40
Fractions / Decimals / Percentages	1, 3, 11, 20, 32, 33	4, 8, 10, 15, 39
Amounts / Proportion	10, 14, 15, 23, 24	25, 37, 38, 40
Mean / Mode	34, 35	
Median / Range	4, 36	
Metric / Imperial Measures / Conversion	9, 29, 30, 37, 38, 40	12, 13, 15, 22, 25,
Currency	7, 10, 11, 21, 28	1, 5, 14, 15, 20, 23
Time / Temperature	5, 6, 16, 17, 36	9, 10, 23, 33, 34, 35
Weight / Capacity	12, 19, 25, 37, 38	30
Area / Perimeter / Volume	13, 18, 24, 25, 26, 27, 29	
Ratio / Scale	2, 8, 22	
Levels of Accuracy / Estimation	9, 28, 32, 38	16, 27
Formulae	24, 27	

Description / Key Coverage: Paper 10	In Question No.	Also in Q No.
Tables / Charts / Graphs	1, 4, 8, 11, 12, 13, 14, 22, 23	16, 17, 18, 20, 28, 30, 33
Fractions / Decimals / Percentages	3, 5, 6, 9, 27, 28, 30, 38	15, 22
Amounts / Proportion	4, 7, 16, 17, 18, 20, 29, 40	30
Mean / Mode	14, 23, 25	
Median / Range	24, 26	
Metric / Imperial Measures / Conversion	31, 36, 39	11, 12, 40
Currency	15, 19, 21, 33, 34, 38	3, 8, 9, 17, 18, 29
Time / Temperature	1, 2, 11, 12, 13	28
Weight / Capacity	4, 32, 36	
Area / Perimeter / Volume	34, 35, 37	
Ratio / Scale	10, 31, 39	11, 12, 13, 17, 33
Levels of Accuracy / Estimation	14, 22	10, 13, 14, 15, 29
Formulae	37	

Description / Key Coverage: Paper 11	In Question No.	Also in Q No.
Tables / Charts / Graphs	1, 2, 3, 7, 17, 19, 26, 39	30, 35, 40
Fractions / Decimals / Percentages	8, 9, 20, 26, 38	22, 23, 25
Amounts / Proportion	10, 13, 25, 26, 36, 35, 40	
Mean / Mode	30, 32	
Median / Range	31, 33	
Metric / Imperial Measures / Conversion	12, 16, 18, 27, 29	21, 22, 23, 24, 37
Currency	4, 28	2, 3, 13, 16, 24, 26
Time / Temperature	1, 5, 7	26
Weight / Capacity	11, 14, 18, 19, 23, 24, 25	20
Area / Perimeter / Volume	12, 21, 22, 27, 37	
Ratio / Scale	6, 23, 29, 34	2, 17
Levels of Accuracy / Estimation	6, 12, 16, 18, 22, 26	5, 11, 14, 37
Formulae	12, 15, 22, 27, 37	

Description / Key Coverage: Paper 12	In Question No.	Also in Q No.
Tables / Charts / Graphs	1, 10, 15, 22, 28, 33, 34	35, 36, 39
Fractions / Decimals / Percentages	5, 8, 9, 23, 24, 25, 26, 39	22, 33
Amounts / Proportion	3, 4, 15, 21, 35, 36	
Mean / Mode	10, 11	
Median / Range	2, 40	
Metric / Imperial Measures / Conversion	7, 14, 19, 21, 22, 31, 32, 37	10, 13, 16, 18, 38
Currency	15, 20, 21	5, 27
Time / Temperature	6, 30, 40	1, 3, 4
Weight / Capacity	13, 14, 18, 19, 29, 31, 32	
Area / Perimeter / Volume	17, 18, 37, 38	
Ratio / Scale	1, 16, 27	
Levels of Accuracy / Estimation	7, 12, 17, 18, 29	35, 36
Formulae	17, 37	

Answers

MULTIPLE-CHOICE QUESTIONS: AoN LEVEL 2

Question	Answer	Question	Answer	Question	Answer
		PAPER ONE			
1	D	2	D	3	B
4	C	5	D	6	C
7	C	8	C	9	C
10	A	11	B	12	B
13	C	14	D	15	D
16	B	17	D	18	A
19	D	20	C	21	D
22	B	23	D	24	C
25	C	26	C	27	C
28	C	29	D	30	D
31	B	32	C	33	A
34	A	35	B	36	C
37	A	38	D	39	B
40	B				
		PAPER TWO			
1	B	2	A	3	C
4	C	5	B	6	D
7	C	8	B	9	B
10	D	11	C	12	B
13	B	14	C	15	D
16	B	17	A	18	C
19	A	20	B	21	C
22	B	23	B	24	C
25	C	26	D	27	B
28	B	29	A	30	A
31	A	32	B	33	D
34	C	35	C	36	B
37	C	38	C	39	B
40	A				
		PAPER THREE			
1	B	2	B	3	D
4	A	5	D	6	C
7	D	8	A	9	B
10	C	11	C	12	D
13	D	14	C	15	D
16	C	17	B	18	A
19	C	20	A	21	D
22	B	23	A	24	B
25	D	26	C	27	C
28	D	29	D	30	C
31	B	32	D	33	D
34	D	35	B	36	A
37	C	38	D	39	B
40	B				
		PAPER FOUR			
1	D	2	A	3	B
4	C	5	D	6	C
7	D	8	D	9	B
10	A	11	D	12	B
13	C	14	B	15	D
16	C	17	A	18	D
19	B	20	D	21	C
22	A	23	D	24	D
25	C	26	B	27	C
28	C	29	D	30	B
31	D	32	C	33	A
34	B	35	D	36	C
37	B	38	C	39	C
40	B				

MULTIPLE-CHOICE QUESTIONS: AoN LEVEL 2

Question	Answer	Question	Answer	Question	Answer
			PAPER FIVE		
1	B	2	A	3	B
4	B	5	D	6	C
7	B	8	A	9	D
10	A	11	B	12	D
13	A	14	C	15	C
16	B	17	C	18	D
19	B	20	C	21	A
22	B	23	C	24	C
25	A	26	B	27	C
28	D	29	A	30	C
31	C	32	B	33	D
34	B	35	A	36	D
37	C	38	D	39	B
40	C				
			PAPER SIX		
1	C	2	D	3	A
4	D	5	B	6	B
7	C	8	B	9	D
10	B	11	A	12	C
13	B	14	D	15	D
16	B	17	C	18	C
19	B	20	A	21	C
22	D	23	C	24	A
25	C	26	B	27	A
28	C	29	D	30	C
31	B	32	C	33	A
34	C	35	C	36	B
37	A	38	D	39	C
40	B				
			PAPER SEVEN		
1	B	2	C	3	B
4	A	5	D	6	C
7	A	8	A	9	A
10	D	11	C	12	D
13	B	14	A	15	C
16	D	17	B	18	C
19	A	20	B	21	D
22	A	23	D	24	C
25	A	26	D	27	B
28	D	29	B	30	C
31	A	32	C	33	B
34	D	35	C	36	C
37	B	38	D	39	B
40	A				
			PAPER EIGHT		
1	B	2	A	3	D
4	B	5	C	6	D
7	B	8	B	9	C
10	A	11	C	12	D
13	D	14	B	15	D
16	B	17	B	18	C
19	A	20	D	21	A
22	D	23	B	24	C
25	D	26	C	27	D
28	B	29	A	30	D
31	B	32	D	33	B
34	C	35	B	36	D
37	A	38	C	39	B
40	D				

MULTIPLE-CHOICE QUESTIONS: AoN LEVEL 2

Question	Answer	Question	Answer	Question	Answer
PAPER NINE					
1	B	2	D	3	B
4	C	5	C	6	B
7	B	8	A	9	D
10	C	11	A	12	D
13	B	14	C	15	A
16	B	17	A	18	D
19	B	20	D	21	A
22	B	23	C	24	A
25	B	26	B	27	A
28	B	29	D	30	C
31	B	32	C	33	C
34	B	35	A	36	C
37	B	38	A	39	B
40	D				
PAPER TEN					
1	B	2	A	3	B
4	C	5	A	6	C
7	B	8	C	9	B
10	A	11	B	12	C
13	D	14	A	15	B
16	A	17	C	18	B
19	D	20	B	21	D
22	B	23	D	24	C
25	A	26	B	27	A
28	C	29	B	30	D
31	B	32	D	33	B
34	C	35	D	36	A
37	C	38	B	39	D
40	C				
PAPER ELEVEN					
1	D	2	C	3	B
4	C	5	A	6	B
7	C	8	D	9	C
10	A	11	B	12	B
13	D	14	C	15	B
16	A	17	C	18	A
19	D	20	B	21	C
22	B	23	A	24	C
25	B	26	D	27	C
28	D	29	A	30	C
31	D	32	A	33	D
34	C	35	B	36	C
37	B	38	A	39	B
40	C				
TWELVE					
1	B	2	C	3	A
4	D	5	C	6	D
7	B	8	C	9	B
10	A	11	B	12	A
13	C	14	B	15	A
16	B	17	C	18	A
19	C	20	B	21	D
22	C	23	A	24	B
25	C	26	A	27	D
28	C	29	B	30	B
31	C	32	A	33	B
34	C	35	B	36	D
37	A	38	C	39	B
40	A				

ANSWER GRID

Name .... Date ....

Instructions

Select one answer choice for each question.

Mark your chosen letter answer with a horizontal line in pencil.

If you wish to change an answer to a question, erase your first choice and select another letter.

1 [a] [b] [c] [d]	2 [a] [b] [c] [d]	3 [a] [b] [c] [d]	4 [a] [b] [c] [d]	5 [a] [b] [c] [d]
6 [a] [b] [c] [d]	7 [a] [b] [c] [d]	8 [a] [b] [c] [d]	9 [a] [b] [c] [d]	10 [a] [b] [c] [d]
11 [a] [b] [c] [d]	12 [a] [b] [c] [d]	13 [a] [b] [c] [d]	14 [a] [b] [c] [d]	15 [a] [b] [c] [d]
16 [a] [b] [c] [d]	17 [a] [b] [c] [d]	18 [a] [b] [c] [d]	19 [a] [b] [c] [d]	20 [a] [b] [c] [d]
21 [a] [b] [c] [d]	22 [a] [b] [c] [d]	23 [a] [b] [c] [d]	24 [a] [b] [c] [d]	25 [a] [b] [c] [d]
26 [a] [b] [c] [d]	27 [a] [b] [c] [d]	28 [a] [b] [c] [d]	29 [a] [b] [c] [d]	30 [a] [b] [c] [d]
31 [a] [b] [c] [d]	32 [a] [b] [c] [d]	33 [a] [b] [c] [d]	34 [a] [b] [c] [d]	35 [a] [b] [c] [d]
36 [a] [b] [c] [d]	37 [a] [b] [c] [d]	38 [a] [b] [c] [d]	39 [a] [b] [c] [d]	40 [a] [b] [c] [d

Lightning Source UK Ltd.
Milton Keynes UK
UKOW07f1924170117

292300UK00014B/725/P

9 781904 995517